34.50
11.95p

PARTICLES AND PARADOXES

PARTICLES AND PARADOXES

The Limits of Quantum Logic

PETER GIBBINS

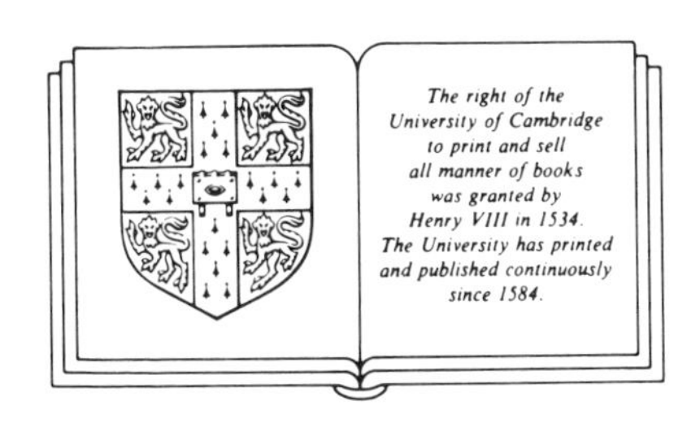

CAMBRIDGE UNIVERSITY PRESS

Cambridge

New York   New Rochelle   Melbourne   Sydney

Published by the Press Syndicate of the University of Cambridge
The Pitt Building, Trumpington Street, Cambridge CB2 1RP
32 East 57th Street, New York, NY 10022, USA
10 Stamford Road, Oakleigh, Melbourne 3166, Australia

First published 1987

Printed in the United States of America

Library of Congress Cataloging-in-Publication Data
Gibbins, Peter.
Particles and paradoxes.
Bibliography: p.
1. Quantum theory – Philosophy. 2. Physics – Philosophy. I. Title.
QC174.12.G52 1987 530.1′2 86–31758

British Library Cataloguing in Publication Data
Gibbins, Peter
Particles and paradoxes : the limits of quantum logic.
1. Qauntum theory
I. Title
530.1′2 QC174.12

ISBN 0 521 33498 5 hard covers
ISBN 0 521 33691 0 paperback

To
Bella and Deb

Contents

Preface — *page* ix

1. Meta-physics — 1

PART I

2. Quantum mechanics for natural philosophers (I) — 19
3. Wave–particle duality — 36
4. The Copenhagen interpretation (I) — 47
5. The Copenhagen interpretation (II): Einstein versus Bohr — 62

PART II

6. Quantum mechanics for natural philosophers (II) — 87
7. Projection postulates — 102
8. Nonlocality and hidden variables — 116
9. A user-friendly quantum logic — 126
10. Quantum logic: what it can and can't do — 142

Conclusion — 166

Notes — 168

References — 175

Index — 179

Preface

Niels Bohr solved the problem of interpreting quantum mechanics once and for all. That, until recently, was the orthodoxy in physics, especially among writers of textbooks on quantum mechanics, who mostly wanted to get off the philosophy as quickly as possible and on with the physics. Niels Bohr, incidentally, would not have agreed with them.

Now the fashion is for physicists to say that quantum mechanics is so peculiar that no one understands it, least of all themselves. Perhaps the change was wrought by Bell's theorem in the 1960s; perhaps it really was just a matter of fashion.

Of course, in saying that quantum mechanics is incomprehensible, one is not saying that it is false, only that the human mind is not tuned in to the way the world is. The philosophy of quantum mechanics deals with the question: What *is* the way the world is, *if* quantum mechanics is true? It would be nice if, in answering that question, one were also to make quantum mechanics comprehensible, something philosophers tend to feel they can pull off.

Philosophers often have another motive, one which is inspired by their intellectual cussedness. Quantum mechanics is most easily interpreted *antirealistically*, that is, as a theory which, though it works, does not describe the way the world is. Therefore, philosophers go out of their way to interpret it *realistically*. Realism in the philosophy of quantum mechanics means the idea that quantum systems are really like classical particles. Everything points against it. Interestingly, the opposite situation occurs with classical physics, which is naturally interpreted realistically, meaning here 'directly corresponds to the world'. So positivist philosophers try to interpret it as a theory about (say) sense data.

'The philosophy of quantum mechanics' is of course an embarrassingly pretentious expression. Perhaps it makes one blush because philosophers are expected to be wise. The philosophy of quantum mechanics, on the other hand, is a bundle of problems.

The central problem in the bundle is this: Quantum mechanics accounts perfectly well for the results of measurements made on quantum systems but it is quite unclear what, if anything, the theory entails about quantum systems upon which no measurement has just been made. Quantum theory gives us a highly workable algorithm for making predictions about the results of measurements, but philosophers and physicists are in total disagreement about what, again if anything, quantum theory tells us about the way the quantum world is.

So does the philosophy of quantum mechanics matter?

It matters because we do physics in order to understand what is true of the world as well as to gain some small (though possibly for us, fatal) mastery over it. So physics and philosophy are not wholly separate and distinct activities. The philosophy of quantum mechanics shows us just how entangled they really are. There is a lesson here even for the unphilosophical physicist. The late J. M. Jauch, certainly a physicist rather than a philosopher, wrote:

> The pragmatic tendency of modern research has often obscured the difference between knowing the usage of a language and understanding the meaning of its concepts. There are many students everywhere who pass their examinations in quantum mechanics with top grades without really understanding what it all means.[1]

Talk of what quantum mechanics asserts about the world, of what is true or false of the world, suggests that we investigate the logic of the language in which we decribe quantum systems. So we explore quantum logic – that typically awkward and peculiar system which drops out so nicely from the formalism of quantum mechanics. In fact, we eventually put quantum logic at the centre of our focus. We use it as a way of looking at recent developments in the philosophy of quantum mechanics. Quantum logic is in any case of great intrinsic interest. There is no better illustration, even within the philosophy of quantum mechanics, of the interconnection of physics and philosophy than the network of problems generated by quantum logic. Questions like: 'Is quantum logic really logic?', 'Is quantum logic a rival to classical logic?', 'Can we speak of a logic of the world?', 'If we can, is this logic to be decided empirically?', 'Can quantum logic be used to resolve the paradoxes of quantum mechanics?', all fall within an area of philosophy which overlaps with the philosophy of logic, with traditional metaphysics, and with physics proper. They are exciting questions, just as exciting to the philosopher of physics as the possibility of non-Euclidean geometries in the theory of space and time.

The idea that this is a classically illogical world is about as deeply metaphysical an idea as one can find. Philosophers naturally gravitate towards the topic of quantum logic, and its employment in attempts to preserve quantum-mechanical realism. Quantum logic and Bell's theorem are the new things in contemporary philosophy of quantum mechanics. For what it's worth, let me say that I think that quantum logic is a logic, and that quantum logical realism fails.

No text as short as this one could begin to handle all the problems of the subject. But I should at least like to give the student enough background knowledge to enable him or her to explore the contemporary literature. In fact, the book has the appropriately incompatible aims of being both an introductory textbook on the philosophy of quantum mechanics and also an essay on quantum logic.

If this book has a single thesis, it is: the lasting results of the philosophy of quantum mechanics are largely negative; quantum systems are not like classical particles; this is not a world of separate things whose behaviour depends only on the way the world nearby is; there is no reason to believe that something like a new classical mechanics can underpin quantum mechanics even when this mechanics is suitably warped by a nonclassical logic. If we are finally left with a mystery, as I think we are, we can still learn from the philosophy of quantum mechanics just how odd the physical world must be.

The book is based on a course of lectures entitled 'Particles and Paradoxes: An Introduction to the Philosophy of Quantum Mechanics' given during the Hilary Term of 1984 at the sub-Faculty of Philosophy at Oxford University, and repeated at the Department of Philosophy of Bristol University in the following year. John Lucas and Rom Harré at Oxford, and David Hirschmann, Adam Morton, and Ian Thompson at Bristol kept me on my toes. My thanks also go to Peter Alexander, Mike Berry, and John Mayberry who read and commented on parts of the text. The many mistakes that have survived their scrutiny are of course my sole responsibility. Finally, thanks must go to my wife Deborah who so tirelessly put up with my strange physicophilosophizing.

Peter Gibbins
Bristol 1986

1

Meta-physics

Metaphysics in the grand style, the product of a philosopher of genius working out from scratch and from his armchair a new conception of ultimate reality, has been out of fashion for a hundred years or so. This is not only because the results of this activity have so often been absurd. It is partly because other disciplines, such as physics, have developed remarkable and puzzling pictures of bits of the world, pictures that are at least as interesting as, at least as 'deep' as, and much more reliable than those of the armchair metaphysician.

Unlike armchair metaphysics, the theories of physics are the results of the efforts of many minds. They are also tested against the world in experiment and are continuously applied in technological devices and in weaponry. We take them seriously. Just why, and in precisely what respects, we should take them seriously is a problem for the philosophy of science. But we do, and if we were asked what is our best account of the way the world is, most of us would cite the fundamental theories of physics.

If armchair metaphysics is out of date, a new kind of metaphysics, scientific metaphysics, has come into fashion. The new metaphysician asks: what is there in the physical world, and what is true of what there is in the physical world? The answers are provided by the philosophy of physics, a subject whose metaphysical part sets out to tell us the way the world is, if physics is true.

Granted that there is a subject called the philosophy of physics, or, as one might say, *meta-physics,* why is there a subject called *the philosophy of quantum mechanics?* And why does it have so much metaphysical interest?

One must of course admit that through nuclear weaponry, the transistor, and now the microchip, our system of communication, indeed our technoculture as a whole, has come to be based on and threatened by an ultimately quantum-mechanical technology. But the importance of quantum mechanics in our scientific and technical culture is not sufficient to generate a *philosophy* of quantum mechanics.

Nor is it sufficient that quantum mechanics is a *fundamental* theory of physics which, incidentally, in its nonrelativistic form it certainly isn't. It is true that quantum mechanics in its nonrelativistic form is our best theory of atoms, molecules, and the solid state, and in relativistic forms, of subatomic particles and matter in the plasma state. Maxwell's electromagnetic theory is a fundamental theory of physics. Yet no one speaks of 'the philosophy of electromagnetism'. Classical mechanics has its philosophy, and important figures in the history of the philosophy of science, philosopher–physicists like Mach and Hertz, have made contributions to it. But 'the philosophy of classical mechanics' has lost much of its significance now that classical mechanics has come to be seen as less than fundamental.

The fact that there is a philosophy of quantum mechanics does not imply the *truth* of quantum mechanics. One cannot say that the quantum theories are known to be true, whereas other less philosophically interesting theories of physics are known to be false. Quantum mechanics is not known to be true. It is a truism of the philosophy of science that no generally applicable physical theory ever could be. In fact one can be sceptical of the truth of quantum mechanics and still be interested in the philosophy of quantum mechanics. One might even have more confidence in the truth of Maxwell's theory than in the truth of quantum mechanics and still think that there needs to be a philosophy of quantum mechanics but no philosophy of electromagnetism.

So why the interest in the philosophy of quantum mechanics?

From the point of view of the natural philosopher, to use an old name for the new metaphysician, today's hybrid of physicist and metaphysician, the most significant difference between the classical and quantum theories is this: quantum mechanics, like electromagnetic theory, is enormously successful in explaining the structure and properties of matter but, unlike electromagnetic theory, it is *deeply* mysterious. It is mysterious because it subverts the classical picture of the world, of which classical mechanics and electromagnetic theory are refinements.

Just how far quantum mechanics subverts the classical picture of the world – which is, crudely, the synthesis of classical mechanics and electromagnetic theory – is a matter of controversy. It is also the subject of this book. Determining just how far quantum mechanics subverts the classical picture of the world is an important part of the philosophy of quantum mechanics. Quantum mechanics seems to contradict atomism, which is strange indeed. It seems to show that things which we think of as separate are in fact not, to show that the world is neither determined nor, perhaps, even determinate. Some physicists and philosophers think that quantum mechanics is so paradoxical to the classical mind that it shows that this is an *illogical world.* This idea – the idea that quantum mechanics has a logic of its own which, if the theory is true, is also the *logic of the world* – is what will concern us most.

The most surprising ideas of contemporary metaphysics have their origin and justification in theoretical physics. This explains why metaphysics is so interesting. The most surprising, maybe the wildest of these ideas are to be found in quantum mechanics. No theory of physics has greater metaphysical interest than quantum mechanics, not even the theories of relativity. No theory has had a more devastating impact on our world-picture. No idea in the philosophy of quantum mechanics is more deeply metaphysical than the idea that the world has a logic and that the logic of the world is quantum logic and not classical logic.

The classical world picture

To fill out the claim that quantum mechanics is fundamental to our metaphysics we should begin by contrasting the world pictures we inherit from classical physics – the physics of Newton and Maxwell above all – with the description of the world that makes up the settled part of contemporary physics.

The *Newtonian world picture,* a refinement of Democritean atomism, is now the common sense of (and so part of the subconscious metaphysics of) most educated laymen and therefore of most professional philosophers. Both materialism and mechanism are embedded in the Newtonian world picture and few ideas can have had greater philosophical influence. The great success of Newtonian mechanics in describing the motions of the planets, the tides, and the behaviour of things on the surface of the Earth in the years following the publication of the *Principia Mathematica* in 1687, gave Newton and his corpuscular philosophy immense prestige.

The corpuscular ontology of Robert Boyle and John Locke, both contemporaries of Newton, was materialist at least as far as the *physical* world was concerned. As a theory about the physical world, it was also mechanistic and deterministic. Newton added forces to this purely corpuscular ontology. God also had a part to play, though Newton's account of it is unlikely to impress us nowadays. Forces enabled Newton to develop his dynamics but only at the cost of admitting what seemed to both Boyle and to Locke along with conventional opinion to be the 'occult'. Newton himself had doubts about forces, doubts which he expressed in the General Scolium of Book III of *Principia,* and at the end of Book III of *The Opticks.*

Of course, no intellectual revolution goes smoothly and it wasn't all plain sailing for Newton. One caricatures the development of the classical world picture if one ignores the opposed views of the Cartesians, of Leibniz, of philosophical (and theological) critics like Berkeley and, later, of the alternative tradition of German idealism. That is, if one ignores the history of philosophy. But the corpuscular philosophy did come to dominate the popular metaphysical outlook of the nineteen century, and Newton's own theological conjectures came to be viewed as an aberrant and antiquated appendage to it.

In the materialist, mechanistic world view which Newtonian physics thus inspired, the physical world was thought of as consisting of enduring particles each of which had determined properties: mass, position and velocity in Absolute Space, that all-pervading medium which defined absolute acceleration and which was perhaps the visual field of the Deity (who unlike us sees things as they are, immediately, without perspective, without time, rather in the way that Cubist painters tried to capture). The force laws which operated between the corpuscles in a system isolated from outside influences specified for that system an unique evolution in time.

The materialism of the Newtonian picture (together with some typically Newtonian theology) is summed up in this delightful and famous passage from Newton's *Opticks,* the book in which his corpuscular philosophy is most clearly expounded.

> All these things being consider'd, it seems probable to me, that God in the Beginning form'd Matter in solid, massy, hard, impenetrable, moveable Particles, of such Sizes and Figures, and with such other Properties, and in Proportion to Space, as most conduced to the End for which he form'd them: and that these primitive Particles being Solids, are incomparably harder than any porous Bodies compounded of them: even so very hard, as never to wear or break in pieces; no ordinary Power being able to divide what God himself made one in the first Creation . . . And therefore that Nature may be lasting, the Changes of corporeal Things are to be placed only in the various Separations and new Associations and Motions of the permanent Particles.[1]

The world view which classical mechanics, as Newtonian mechanics in its nineteenth-century form came to be called, was both materialist and determinist and at least potentially a-theist. The most famous expression of the thesis of universal determinism is to be found in the second chapter of the *Philosophical Essay on Probabilities* written by the Marquis de Laplace in 1816:

> We ought then to regard the present state of the universe as the effect of its anterior state and as the cause of the one that is to follow. Given for one instant

an intelligence which could comprehend all the forces by which nature is animated and the respective situation of all the beings who compose it – an intelligence sufficiently vast to submit these data to analysis – it would embrace in the same formula the movements of the greatest bodies of the universe and those of the lightest atom; for it, nothing would be uncertain and the future, as the past, would be present to its eyes. The human mind offers, in the perfection which it has been able to give to astronomy, a feeble idea of this intelligence. Its discoveries in mechanics and geometry, added to that of universal gravity, have enabled it to comprehend in the same analytical expressions the past and the future states of the world.[2]

In what we shall call *classical metaphysics* – the metaphysics naturally appended to classical mechanics – the physical world is *determinate:* things consist of separate corpuscles each of which endures through time, the variable physical quantities or *dynamical variables* used to describe bodies – energy, momentum, angular momentum – all take a continuous range of possible values, each possible value being definite, a point on the real line, or a point vector in a three-dimensional space. The machinery of the physical world operates on these corpuscles through the iron laws of conservation of energy, momentum, and angular momentum.

The world picture of classical physics in its late nineteenth-century form superimposed on this Newtonian ontology something undreamt of in Newton's time – the electromagnetic field and the aether in which it inhered. The aether in itself was nothing new. In one form or another it appears in the metaphysics of René Descartes and Giordano Bruno and others. But James Clerk Maxwell's electromagnetic field theory, set out in his treatise of 1871, added the *field* to matter and the aether. Physics became dualist. The electromagnetic field explained the apparent *action-at-a-distance* of electricity and magnetism. Solutions to Maxwell's equations showed that the electromagnetic field propagated through space with a fixed speed as radiation in the form of waves. Maxwell had shown that light was nothing more than electromagnetic radiation of a particular range of wavelengths.

A medium to support this wave motion had to be invented. After all, waves have to be waves *in* something. The *aether* supported electromagnetic waves and was perhaps to be identified with Newton's Absolute Space. But this neat encapsulation of mechanics and electricity and magnetism concealed a fundamental tension.

Given the aether hypothesis, the speed of light is its speed relative to the aether. Therefore Maxwell's equations, which contain a constant referring to the speed of light, must lack the universality of Newton's laws of motion, for these were held to be true in all unaccelerated frames of reference and not just those at rest with respect to the aether. The dualism of field and particle was unsettling in itself. Particles are highly localized and impenetrable. Fields and waves in fields are extended in space. Waves can be superposed on one another and even made to self-interfere. Consequently, the prevailing physical picture of the world embodied in the physics of the late nineteenth century was essentially Newtonian but it had, rather untidily, to incorporate a field. Soon the field was to threaten the Newtonian ontology in a fundamental way.

The tension between Newtonian mechanics and Maxwellian electromagnetism produced an unexpected outcome in the special theory of relativity. It was Newtonian mechanics that gave way to electromagnetism. Einstein dropped the aether hypothesis. Maxwell's equations acquired full generality. The Newtonian equations were seen to be mere limits of the relativistic equations (as the speed of light is allowed to go to infinity). Furthermore, the speed of light was seen to be a constant in all allowed frames of reference.

Through the theories of relativity – the special theory of 1905 and the general theory of 1915 – theoretical physics revolutionized our ideas of space and time. These are now thought of as aspects of a single entity called *space-time,* the arena of physical happenings. An oddness of the relativistic picture of the world is that space and time do not fall out of space-time in an unique way. They fall out differently for relatively moving observers. General relativity tells us that space-time is not only an unity but also that it is warped. But deep though the philosophical impact of relativity is, the impact of quantum mechanics is even greater.

In special relativity space-time takes over from space and time is perhaps the aether but in a new guise. In general relativity space-time is warped and this generates what appears to us as the force of gravity. *Relativity – special and general – preserves the determinateness of reality.* In comparison with quantum theory, relativistic physics seems merely neoclassical.

Quantum metaphysics

One should mention the impact that relativity has had on our conception of the world for the following reason. With quantum mechanics we meet something entirely new, something much more subversive of the classical picture of the world even than relativity. Again the revolution took time to unfold. In a first phase, which takes up the period from Planck's explanation of black-body radiation in 1900, through Einstein's explanation of the photoelectric effect in 1905, the Bohr theory of the hydrogen atom in 1913, the Bohr-Sommerfeld theory of the atom of 1916, and the work of the spectroscopists up to 1925, the *old quantum theory* quantized the classical ontology.

But in a second phase, beginning with Heisenberg's matrix mechanics and Schrödinger's wave mechanics in 1925, the classical ontology was swept away. The mathematical formalism of quantum mechanics, the quantum theory of this second period, was developed long before its interpretation was decided. Indeed, one should say that it never has been decided. There is a philosophy of quantum mechanics precisely because there is no agreement as to what the theory tells us about the world.

In what philosophers habitually call 'the philosophy of quantum mechanics' we pursue quantum physics to 1930 or so, up to the time at which the formalism had been unified by the Hungarian mathematician von Neumann, the indeterminacy relations derived and discussed, prior to most of the early work on quantum electrodynamics and quantum field theory.

Why do we say nothing about those relativistic theories, quantum electrodynamics, quantum field theory, and fundamental particle physics which paint pictures of the deepest and most fundamental ontology of physics, if there is such a thing? Why do we limit ourselves to the nonrelativistic quantum mechanics of sixty odd years ago, physics which is surely old hat by now?

First, the essentially quantal aspects of all these theories – like the superposition principle, the indeterminacy relations, complementarity – already appear with quantum mechanics.

Second, although quantum mechanics cannot be true strictly speaking since it is a nonrelativistic theory, quantum field theory contains inconsistencies and there is much greater uncertainty as to its correct formulation than is the case for quantum mechanics.

And third, the philosophy of elementary nonrelativistic quantum mechanics is hard enough already. In fact philosophers are only now beginning to seriously examine quantum field theory.[3]

So we filter out as far as possible the electromagnetic and relativistic aspects of the quantum theories, much as a nineteenth-century philosopher of science like Ernst Mach might filter out electromagnetism and its field from his philosophy of classical mechanics.

What then are the problems that the philosophy of quantum mechanics confronts?

Unlike classical mechanics, quantum mechanics is at least prima facie a *fundamentally statistical* theory, a fact on which much turns. Quantum mechanics yields probabilities. Only in special cases can it make determinate nonstatistical predictions. Contrast classical particle mechanics. Every particle in a system described by that theory has a determinate position and momentum at any given time. The force laws that operate on the particles determine the way these positions and momenta change with time. When the number of particles is large – in fact when the particles interact any number greater than two is large – the physicist must fall back on a statistical description of the system. The stock example is provided by the kinetic theory of gases. One can treat a sample of a gas as a collection of classical particles and one can predict its gross behaviour even though one cannot in fact predict the behaviour of any particle in the sample. One says that one predicts the behaviour of each particle *in principle* though not in fact since one knows the force law governing the interactions and one knows that these fix an unique evolution in time for the position and momentum of each particle.

Probability – as it appears via the Maxwellian distribution function for particle speeds in a gas – arises because of our *ignorance* of the state of each particle in the sample of gas. It is the same throughout statistical mechanics and throughout classical physics generally. The classical world picture has it that the physical world is determinate and determined and that probability is not a feature of the world but arises because of our ignorance and so is a feature of us, or of the relation in which we stand to the world as knowing subjects.

If the probability in statistical mechanics arises because of ignorance, then one must ask: what do probability statements mean? What do we mean when we say that the most probable speed for a nitrogen molecule in a sample of air at room temperature is 420 meters per second?

The standard interpretation of probability in physics has it that probability deals with *relative frequencies within an ensemble*. The probability that our molecule has its speed within a given range is then just the fraction of molecules having speed in that range in the appropriate ensemble. The appropriate ensemble might be just the actual collection of nitrogen molecules in the sample of air. Understood this way probability has nothing to do with the individual molecule but is simply a property of the ensemble. When we say that the probability that a particular molecule has its speed in such-and-such a range is p we can legitimately mean only that the fraction of molecules in our chosen ensemble having speeds in that range is p. We reduce probability statements about the molecules in a gas to statements about relative frequencies in ensembles, all the while conceding that the motion of each individual molecule is determinate and rigidly determined by the laws of classical mechanics.

But quantum mechanics is, or at least seems to be, irreducibly statistical. What does this mean? It means that quantum mechanics is unlike statistical mechanics in that it is not and cannot be based on, underpinned by, a deeper nonstatistical mechanics. So should we say that quantum mechanics *does not describe individual quantum systems at all* but describes *mere ensembles* of quantum systems because probability refers to ensembles? Should we say that the quantum theories fail in their duty to provide a determinate and deterministic underpinning to the probability statements they issue?

Einstein thought so. He became a critic of the orthodox interpretation – the Copenhagen interpretation – of quantum mechanics as it developed and took hold in the late 1920s, somewhat ironically since Einstein was one of the founders of the old quantum theory and even received his Nobel Prize not for relativity but for his work on the photoelectric effect. According to the Copenhagen interpretation quantum mechanics exhaustively describes the individual quantum system, the individual electron, proton, photon. Einstein, on the other hand, held that quantum mechanics does not describe the individual quantum system for which variables 'hidden' from quantum mechanics would be required, though it does describe the gross statistical behaviour of ensembles.

The hidden-variables idea has it that there really is a (presumably deterministic) underpinning of quantum mechanics analogous to the underpinning of classical statistical mechanics by classical mechanics. The ensemble view of quantum mechanics seems to leave open the route to a hidden-variables theory. It views quantum mechanics as a new and generalized statistical mechanics.

From the fact that quantum mechanics is a statistical theory and from the relative frequency interpretation of probability arise two of the most fundamental questions in the philosophy of quantum mechanics: *Does quantum mechanics describe only ensembles or does it describe the individual quantum system? Can there be a hidden-variables underpinning of quantum mechanics?*

Taken together these questions form the problem of the *completeness* of quantum mechanics. An ensemble interpretation which is sympathetic to the idea of hidden variables (like Einstein's) will view quantum mechanics as incomplete. An individual-system interpretation (like Heisenberg's) which rejects the possibility of a hidden-variables underpinning will view quantum mechanics as either complete or at least as complete as it is possible for a fundamental theory of physics to be. The dispute between the two interpretations has philosophical significance not merely for the philosophy of physics but in philosophy generally. For in defending the maximal completeness of quantum mechanics one can be led, as Bohr was, to develop a rudimentary philosophy of the limits of explanation and even of the limits of language. The dispute has extraphilosophical significance since someone who rejects the possibility of hidden-variables theories will frown on research into such theories while his opponents will try to encourage it.

Several questions related to completeness arise immediately. There is the question of the meaning of the Heisenberg uncertainty principle – or as we say more neutrally, *the indeterminacy relations*. These tell us about a trade-off between 'indeterminacies' in (for example) position and momentum. Do the indeterminacy relations apply to the individual system or must they be interpreted, perhaps like probabilities, as referring only to ensembles?

Assuming an individual system interpretation, the indeterminacy relation

$$\delta X \delta P_X \geqslant \tfrac{1}{2}\hbar$$

tells us that the product of the 'uncertainties' in the position (in the X-direction) and momentum (in the X-direction) of a quantum system must be greater than or equal to a minimum whose value depends on $\hbar$, Planck's constant. Yet it is not clear what 'uncertainty' is supposed to mean.

Reinterpreting the indeterminacy relation, the ensemble interpreter might say this. Imagine a beam of particles, all prepared by the same process to have their momenta restricted to a narrow range. Imagine for example a beam of nuclei emerging from a mass spectrometer. When you measure the positions of some of the particles and the momenta of some others as they emerge from the preparing apparatus you will find that they have minimum spreads in their positions and momenta such that the product of these spreads will always satisfy the indeterminacy relation for position and momentum. This says nothing at all about whether each individual nucleus has simultaneous determinate position and momentum. Indeterminacy is demystified but at the cost of limiting the scope of quantum mechanics. If one goes on to insist that each individual nucleus has simultaneous determinate position and momentum then one has an *epistemic* interpretation of the indeterminacy relation. Indeterminacy expresses a limitation on our *knowledge* of an unfuzzy Nature.

In individual-system interpretations, like the Copenhagen interpretation as presented by Heisenberg, the indeterminacy relations present more of a problem. They can be thought to express a real fuzziness in Nature: the *ontic* interpretation, asserting a *de re* indeterminacy, an indeterminacy of things not of knowledge; or, alternatively, a limitation on the simultaneous definability of dual or conjugate concepts like position and momentum: asserting a *de dicto* indeterminacy, an inevitable indeterminacy of concepts rather than of things.

The meaning of the indeterminacy relations and the problem of completeness formed the basis of the Bohr/Einstein dialogue, the active core of the early philosophy of quantum mechanics. At first Einstein tried to show that the Copenhagen interpretation of the indeterminacy relations was false, that clever thought-experiments could be devised which circumvented them. But Bohr succeeded in showing to the satisfaction of many, though not to Einstein, that in each of Einstein's examples one could apply the indeterminacy relations consistently if you applied them sufficiently thoroughly.

The dispute about the indeterminacy relations took up the first period of the dialogue, from 1927 until 1930. But one could say that Einstein's real interest, even in this period, lay in showing that quantum mechanics is incomplete. In the second phase Einstein tried to demonstrate incompleteness directly, by considering a thought-experiment and a philosophical argument known as *EPR*: the Einstein–Podolski–Rosen argument. EPR showed that either quantum mechanics is complete or that quantum mechanics implies nonlocality, instantaneous action-at-a-distance. Einstein showed that if it is admitted, as it is by the Copenhagen interpretation in one of its forms, that the act of making a measurement on a quantum system disturbs it, then this disturbance can be transmitted instantaneously over large distances. Einstein rejected action-at-a-distance on principle and so considered that he had demonstrated the incompleteness of quantum mechanics.

EPR reverberates down to the present day. A deeper analysis of EPR due to J.S. Bell in the middle 1960s shows, so most philosophers of physics would say, that quantum mechanics is inconsistent with any hidden-variables theory that rejects action-at-a-distance, and further that quantum mechanics is itself a nonlocal theory. Experiments, though difficult ones to perform, can decide between quantum mechanics and any local hidden-variables theory. The consensus is that experiment has vindicated quantum mechanics and also refuted locality.

EPR is a paradox which belongs in the cluster of puzzles connected with measurement. For an individual-system interpretation, measurement is a problem. When a quantum system is not subject to measurement by an observer its state evolves deterministically in accordance with an equation – Schrödinger's equation – just as the state of a classical particle in classical particle mechanics evolves according to Newton's second law of motion. When someone makes a measurement on it, its state is presumed (in practically all cases) to jump discontinuously and acausally into another state. This acausal feature of measurement in quantum mechanics is often called (picturesquely but perhaps unfortuantely) *the collapse of the wave-packet,* because the particle is associated with a wave which describes its state. In this picture, making a measurement of position amounts to instantaneously collapsing the wave into the region in which the particle is found.

Is the collapse of the wave-packet a physical process? If it is, what distinguishes measurements from other interactions which cause quantum mechanical states to evolve continuously? Is it the fact that there is *consciousness* at the tail of the process? If so, can dogs as well as humans cause the wave-packet to collapse? If the answer to all these questions is 'yes', does quantum mechanics point to the final triumph of idealism over realism, consciousness making an indeterminate world determinate?

Less picturesquely, quantum-mechanical states can be represented as superpositions of states. Take two allowed states of a system, 'add' them, and you have a third allowed state. When one gives a quantum-mechanical account of measurement – the interaction between a quantum system and the measuring apparatus – one finds that the measuring apparatus should generally end up in *no definite state at all*. One finds that the measured system plus measuring apparatus is in a superposition of states corresponding to the superposition in which the measured quantum system found itself. But we have only to look around to see that the world is not in a superposition of states. Superposition does not imply a little fuzziness, it implies that, for example, experimental pointer readings should be (say) both 'up' and 'down', or more exactly, in no single direction at all.

To avoid all these puzzles about measurement it seems we have to limit the application of quantum mechanics somehow. Perhaps there must be a 'cut' between the observer and his apparatus on the one hand (described classically so as to prevent superposition of states in the determinate macroscopic world we inhabit) and the quantum world on the other. But if so where should we put this cut? Does it matter where we put it? Does this mean that classical mechanics is not a limiting case of quantum mechanics in the way in which it is a limiting case of relativistic dynamics? Does quantum mechanics *presuppose* classical mechanics? Must we give up our ideal of giving a single consistent overall picture of the physical world?

One of the more attractive features of the philosophy of quantum mechanics is that most of its problems can be handled via thought-experiments, most of which can be made into real experiments. Two thought-experiments stand out as capturing between them most of the essential puzzles: the two-slit experiment and EPR. The two-slit experiment illustrates *wave–particle duality* and the puzzle of quantum mechanical measurement. EPR illustrates nonlocality, the oddity of quantum-mechanical measurement, in a new way. It also provides a medium for discussing incompleteness. These two thought-experiments, the two-slit experiment and EPR, keep turning up in one form or another.

It is a feature of the paradoxes that what is paradoxical is not the thought-experiment but the description we give of it. The world simply is. It cannot be paradoxical. Therefore the paradoxes – of measurement and nonlocality – encourage interpretations of quantum mechanics which employ a *nonstandard logic*. In the most significant interpretations of this type the nonstandard logic used is *quantum logic*, a structure which drops nicely out of quantum mechanics itself. Quantum logical interpretations of quantum mechanics are a growth area in the subject. But quantum logic has strange nonclassical features. For example, suppose that P is not true and that P or Q is true, we can infer that Q is true in classical logic but not in quantum logic. So several questions arise, questions which will preoccupy us in later chapters of this book. Is quantum logic really logic? Do quantum logical interpretations actually work? That is, do they smooth away the paradoxes? Can we come to a new understanding of the quantum world if we limit the logic of our quantum descriptions to quantum logic? And if we can, is quantum logic worth the price?

These problems, attracting a mix of logic and physics, are especially intriguing to the philosopher. For here the philosophy of physics meets the philosophy of logic and the philosophy of language. Quantum logic is a fascinating, somewhat intransigent formal system, of interest in its own right. The idea that the world may be so strange as to demand a new logic is one of the most profound in metaphysics. Therefore we choose to view quantum logic as the vortex around which our treatment of quantum mechanics revolves and converges.

These then are the main problems of the philosophy of quantum mechanics that confront us:

Does quantum mechanics describe individual quantum systems or only the statistical behaviour of ensembles of quantum systems?

Are quantum systems waves, or particles, both or neither?

How should we understand the uncertainty principle?

What does probability mean in quantum mechanics?

Do individual quantum systems have simultaneous precise values for all their dynamical variables?

Do electrons have trajectories?

Can there be a deterministic, hidden-variables underpinning of quantum mechanics?

How are we to understand measurement in quantum mechanics?

Is quantum mechanics a nonlocal theory?

Are there genuine paradoxes in quantum mechanics, and if so how can they be resolved?

Can they be resolved by restricting our reasoning about individual quantum system to quantum logic?

Is quantum logic a logic?

Is it a rival to classical logic?

The philosophy of quantum mechanics is in the business of interpreting the formalism of the theory, a mathematical object assumed to be given independently of the various possible interpretations, as if formalism and interpretation were entirely separable things. It is broadly true that the formalism preceded its interpretation. But the two naturally went hand in hand. Matrix elements, in Heisenberg's mechanics, and eigenvalues in Schrödinger's wave mechanics were naturally associated with energy levels before the details of the formalism or the details of the interpretation were spelled out. Perhaps it would be better to say that the formalism came partly interpreted and that the philosophy began when it was felt necessary to flesh out that partial interpretation into a picture of the world.

Unfortunately for us philosophers, the fleshing out of the quantum mechanical formalism makes the philosophy of quantum mechanics rather more mathematical than any other branch of the philosophy of science. But there is every reason to keep the mathematics to a minimum. In the next chapter we reveal quantum mechanics to the extent that we can then handle the philosophical ideas of the physicists Schrödinger, Heisenberg, Bohr, and Einstein in the 1920s and 1930s. This is the philosophy of quantum mechanics part I and it takes up Part I of this book.

For Part II we need some logic and some more mathematics in order to treat the more recent philosophy of quantum mechanics. There we discuss hidden-variable theories, nonlocality and quantum logic. We need to hear something about Hilbert space.

The philosophy of quantum mechanics is a complex part of the overlap between philosophy and physics. How can one best approach it?

The line we take consists in examining the clash between proposed interpretation and paradox. Avoidance of paradox is one of the boundary conditions of the subject. Quantum mechanics makes paradoxes for the *realist,* the man who thinks that quantum systems like electrons and photons and protons and neutrons are essentially like classical particles, by virtue of always having precise values for all dynamical variables. This, unfortunately, is what that overworked word *realism* means in the philosophy of quantum mechanics.

Therefore this book is about the way the particle view of the quantum world makes for paradox and how these paradoxes may be dissolved. But the difficulty of the philosophy of quantum mechanics is liable to generate, among philosophers of physics, that self-consciousness about the activity of philosophizing which is such a feature of the analytical tradition from which English-speaking philosophers can hardly expect to escape. Therefore this book is also about the reflections quantum mechanics forces us to make on the nature of theorizing in physics and on the possibilities open to the philosophy of physics.

Whatever its exact philosophical outcome, quantum mechanics teaches us to doubt the simplicities of the Democritean ontology in which we can picture ourselves as living in a world of separate, determinate Atoms in the Void.

What the philosophy of physics is

Of course one cannot do the philosophy of physics without pondering what it is that we can reasonably expect of the subject. For the philosophy of physics is part physics and part philosophy. Its business is understanding physics and, because it is part philosophy and so self-conscious and self-critical, understanding understanding physics. Here a nonterminating loop threatens, as it tends to in philosophy.

Another symptom of the philosophy in the philosophy of physics is a quite justified agonizing about what the subject is and about what it ought to be. If there is a single thought behind what I take to be some of the most interesting developments in recent philosophy of physics – as found in the work of Nancy Cartwright[4] and Ian Hacking[5] for example – it is this: that the philosophy of physics is the philosophy of *real* physics. What else could it be, if it is not to be just another philosopher's fantasy? As corollaries: physics is not what philosophers of science and of physics usually take it to be; the orthodox conception of the philosophy of physics is flawed.

To most philosophers of physics working in the broad analytical tradition it seems that there is a job of analysis to be done. It seems that physics conceals its real representation of the physical world in an unregimented confusion of *physics-as-done-by-physicists*. Physicists use the notions of force, particle, and field and so some theory of physics may appear to require forces, point-masses, and continuous and pervasive fields. But do these ideas correspond to something in the world or are they parts of the representation that physicists make of the world which correspond to nothing? In the case of quantum mechanics such questions become crucial. Could it be that though physicists speak of individual particles there are no such things? Could it be that the logic of the world is not the logic physicists use in doing physics? Could it be that the paradoxes of quantum mechanics are no such thing and are merely the side-effects of using the wrong logic?

There is an inevitable tension between the acceptance of physics by the natural philosopher and his reforming it, or at least reforming the way physicists habitually present it. The philosophy of real physics derives whatever interest it has from the presumed success of physics. It is parasitic on physics. Yet the reforming natural philosopher sets out to reconstruct physics, or at least to extract from physics that representation of the world that physics should be seen as yielding. There is a problem, hardly ever addressed, as to whether this should be attempted. There is also a problem about how it should be attempted, if it should at all.

To this second question the orthodox, or 'theoreticist' view of the subject, has an answer, an answer which is composed of a battery of ideas which make up an account of what the philosophy of physics is. The philosopher of physics should reveal what the theories of physics are, or must be committed to, via *a logical analysis and reconstruction* of the representations of that world that physicists make. The representations are *theories of physics* and the process of analysis necessarily results in the exhibition of the theories of physics as mathematical apparatus plus bridge principles which give the mathematical skeleton some empirical flesh. Theories are what interest the theoreticist philosopher of physics. In fact, *physics is a set of theories ready for logical analysis.* A theory of physics is a set of propositions, and so an appropriate object for the logician's attention.

The orthodox, theoreticist view of physics combines a reaction to logical empiricism (in the proposition that a scientific metaphysics is possible) with a respect for the formal methods and the tools of logical analysis that the logical empiricists and their successors developed. Blending as it does these separate influences from physics, philosophy, and logic, theoreticism has the power to make the philosophy of physics interesting to philosophers.

This book is about what I take to be a failure in pursuit of the theoreticist's ideal, though the ideal itself is a worthy and defensible one. That theoreticism itself is defensible is one of the subthemes of this book. But quantum mechanics resists the logician's mollification. It cannot be domesticated by logic. Its principal philosophical outcome (I suggest, humbly following Niels Bohr) lies in presenting us with the limits of theorizing in physics, the limits of our power to represent the physical world.

Part I

2

Quantum mechanics for natural philosophers (I)

The quantum revolution began in 1900 with a novel solution to what might seem to be one of the minor puzzles of theoretical physics, the problem of black-body radiation: the problem of the interaction of those two sorts of entity which figure in the dualist ontology of late classical physics.

The trouble for classical physics was that the facts about a black body's capacity to absorb and emit radiation defied thermodynamic explanation. The best that classical physics could do was to get the facts wrong and to generate paradox. So what was the problem of black-body radiation and why was it considered to be so puzzling?

First, what is a black body? Some bodies absorb and emit radiation more readily than others. A mirror is a poor absorber of radiation in the form of light, a piece of coal is a good one. A good absorber of radiation in the form of light is 'blacker' than one that is not.

One can extend the metaphor to radiation outside the visible range. Clearly, a body that is 'blacker' than another for one range of radiation frequencies need not be 'blacker' for all frequencies. A body's 'blackness' will, in general, depend on its chemical structure. But we can imagine a body whose blackness is maximal for all wavelengths. Such a body, if it exists, has a blackness which is independent of its chemical structure. We call it a *black body*. Two interesting questions are: first, are there bodies in Nature which are uniformly and maximally black for all wavelengths? and second, how does such a black body distribute the energy that it radiates as a function of the wavelength of the radiation it emits?

To grasp the classical answers to these questions we need some classical thermodynamics, but not much. A dimensionless quantity (a real number between zero and one) called the *absorptive constant* $A(\nu)$ of a body is defined as the proportion of the electromagnetic radiation incident in the frequency range ν to $\nu + \mathrm{d}\nu$ which it will absorb. Similarly, the *emissive power* of the body at frequency ν is defined as the amount of radiation energy which it emits per unit time. Classical physics predicts that the ratio of the emissive power to the absorptivity is fixed for any given body (black or not) and is independent of the frequency ν. An object whose absorptivity equals unity for all wavelengths is one of our black bodies. A black body is therefore something which emits and absorbs radiation with maximum possible effectiveness. Furthermore, there are such things.

The inside of a cavity made of any opaque material at equilibrium provides an excellent approximation to a black body, so there are indeed such things, at least to a good approximation: think of a piece of radiation (on the quantum picture, a photon) bouncing around inside the cavity for ever, attenuating slowly (or abruptly on the quantum picture), being pretty well perfectly absorbed after a time. Remember that emissivity and absorptivity go hand in hand. So a cavity will be a maximal emitter too.

Now imagine a cubical cavity, a black body in other words, in which all possible waves of electromagnetic radiation exist. At the walls of the cavity the waves must have nodes, which is to say that the wave intensity must go to zero. The number of allowed waves obeying this condition increases as ν increases. More and more waves of smaller wavelength can fit in. But according to classical statistical mechanics – if we identify the cavity nodes with classical 'particles' – each wave can carry the fixed amount of energy kT, where k is a constant and T the absolute temperature.

Calculate the total energy in the cavity and you find that it goes like the integral of ν^2 with respect to ν, an integral which goes like ν^3, with upper limit infinity and lower limit zero. So the energy in the cavity should be infinite. Worse, all the energy will tend to be concentrated in the high-frequency, or large ν, end of the spectrum. The radiation in the cavity should get bluer and bluer. This is the *ultraviolet catastrophe* as the physicist Ehrenfest later called it. No solution to this disastrous outcome was found in the nineteenth century even though the problem was central to classical statistical mechanics. In any case the attempt to describe the radiation in a cavity at equilibrium certainly could not be made to conform to the scrupulously accurate measurements of the experimental physicists Lummer and Pringsheim. The classical explanation naturally got the facts completely wrong.

After some years of puzzlement about the problem of black-body radiation, Max Planck in 1900 partially resolved the paradox by in effect proposing that the exchange of energy between matter and radiation could take place only *discontinuously*, via radiation packets of a restricted size

$$E = nh\nu$$

where n is a positive integer, h is the famous Planck's constant, and ν is the frequency of the radiation exchanged. Planck's quantum hypothesis succeeded in avoiding the ultraviolet catastrophe and in squaring theory with Lummer and Pringsheim's experimental results, Planck's distribution of irradiated energy from a black body against radiation frequency being called *Planck's law*.

Planck himself took what we now think of as a conservative view of the quantum hypothesis: discontinuity might be a feature of energy exchange between matter and field but the exchanged packets of energy, the quanta, need not retain their identity during transmission through space. Continuity was supposed to reign, save in the comparatively exceptional case of the exchange of energy between matter and the electromagnetic field.

In 1905 Einstein solved another outstanding problem of classical physics by proposing an even more radical quantum hypothesis: that there were *free light quanta,* 'photons' as we now say, which did retain their identities while traveling through space. The idea enabled him to explain the classically puzzling photoelectric effect.

If you shine light of the appropriate frequency on a clean piece of an alkali metal like sodium in a vacuum the metal emits electrons. This much is quite reasonable on the classical theory. The details of the photoelectric effect are not. First, there is a threshold frequency below which no electrons are emitted even though the amount of energy shone on the metal surface can be as large as you like. Second, the maximum energy of the emitted electrons depends only on the frequency and not on the total energy of the light incident on the metal surface. Third, emission starts as soon as the light is turned on.

A classical explanation of the photoelectric effect would have to view it as a kind of boiling of electrons from inside the metal. But boiling is not instantaneous and will occur however weak the heat source so long as heat is not allowed to escape. Clearly the boiling picture is all wrong. In Einstein's explanation, the photoelectric effect was not to be thought of as a boiling off of electrons but more like the knocking of snooker balls out of the opening triangle with the cue ball. Light was to be thought of as corpuscular, along Newtonian lines. Each light quantum of frequency ν and wavelength λ (so that $c = \nu\,\lambda$) was to have energy

$$E = h\nu$$

and momentum

$$p = h/\lambda$$

Black-body radiation and the photoelectric effect were early though controversial successes for the quantum hypothesis. The community of physicists felt satisfied with neither. Planck's hypothesis appeared ad hoc. Einstein's explanation of the photoelectric effect remained controversial until the mid-1920s. In fact Niels Bohr, no conservative himself, was for long one of the staunchest opponents of the free light quantum.

A further partial success for the application of quantum ideas lay in Einstein's explanation in 1908 of the specific heats of solids at low temperatures. But the most remarkable achievement of this period of quantum theory – which is called *the old quantum theory* – was Niels Bohr's theory of the hydrogen atom of 1913.

William Rutherford had interpreted his alpha-particle scattering experiments as showing that atoms consisted of a small positively charged central core and an outer arrangement of negatively charged electrons. Most of Rutherford's alpha-particles (or helium nuclei) passed straight through his metal foils with very little deflection. But a small fraction suffered large deflections and some were turned around through 180 degrees. Both facts are consistent with the idea that an atom consists of a small central positively charged nucleus surrounded by a planetary system of outer electrons.

The puzzle for classical physics is this: what stops the atom collapsing and what explains the stability of matter? If the electrons are stationary then Maxwellian electromagnetism tells us that they will be attracted to the central core. If they rotate they should, as charged particles accelerated in an electric field, radiate their energy and (quickly) collapse into the core.

Bohr's apparently ad hoc solution to the puzzle was rather like Planck's for black-body radiation: the electrons in an atom (in the hydrogen atom in particular) are to be allowed to be stable only in certain classically possible orbits, in those orbits in which the angular momentum of the electron is an integral multiple of h, Planck's constant. Classically the electron should go spiralling down towards the nucleus very rapidly, say in 10^{-10} sec or so. But Bohr simply forbade this, boldly (as Popperians say).

From this idea, which fixed the energy levels of the electron in the hydrogen atom, together with the idea that the wavelengths of radiation emitted and absorbed by the atom were due to transition between the allowed stable orbits, the spectrum of hydrogen could be accurately inferred. This was a great, perhaps the great triumph for Bohr and the old quantum theory. The extension to more complex atoms of Bohr's idea of quantizing electronic orbits resulted in the partially successful *Bohr–Sommerfeld* theory of the atom.

Unfortunately for the old quantum theory, the ad hoc attempt to graft quantum conditions onto a classical ontology worked only for the hydrogen atom. The years from 1913 to 1925 saw a number of successes, in which Einstein's semiclassical quantum theory of radiation of 1917 is perhaps the most important. But they were also the years when research in atomic physics was merely 'systematic guessing, guided by the correspondence principle',[1] a period of the history of physics brought beautifully to light in Russell McCormmach's fine novel *Night Thoughts of a Classical Physicist.*[2]

The correspondence principle, Bohr's most important methodological guideline, tells us that the numbers given out by a classical and a quantal explanation of some phenomenon will converge as the phenomenon 'gets bigger', as the quantum numbers get larger, as in other words the phenomenon gets less specifically quantal. Quantum mechanics proper itself grew out of the old quantum theory. It appeared in two very different forms in the years 1925 and 1926. Heisenberg's austere matrix mechanics developed from a mathematical formulation of the correspondence principle. Schrödinger's wave mechanics grew, under the influence of de Broglie's matter/wave idea.

In the meantime the duality of wave and particle had become a real problem for physics. Few physicists had taken Einstein's idea of the free light quantum seriously. But in 1922 the American physicist A. H. Compton calculated that on the basis of the particle model of light, an X-ray scattered from a free electron should undergo a change in its wavelength, something which would be inexplicable on the classical wave theory. Compton performed the experiment and found what was predicted by the 'light as billiard ball' model. Compton was awarded the Nobel Prize in 1927 for his discovery. From then on, wave–particle duality was a controversial feature of physics. The discovery of the Compton effect was a turning point in twentieth century physics, as Roger Stuewer's excellent book *The Compton Effect*[3] suggests.

The French physicist (and eccentric) Louis de Broglie proposed in a series of papers from 1923, and in his famous thesis of 1924, that matter is wavelike, thus inverting Einstein's assertion that light is corpuscular. De Broglie's theory was relativistic. Every particle, every electron, proton and so on was to have an associated wave, of frequency ν_0 given by

$$m_0 c_2 = h\nu_0$$

when at rest, where m_0 is the rest mass of the particle. When moving, a particle with momentum p has a de Broglie wavelength λ such that

$$p = h/\lambda.$$

Here, λ is clearly the wavelength of a monochromatic, a pure sine wave.

More accurately in de Broglie's account, a localized particle is associated with a superposition, or group, of waves. Its velocity is the *group* velocity of these waves. The velocity of each of the waves in the group is greater than the speed of light, and so the original de Broglie wave of wavelength λ cannot carry energy and cannot be associated with the original particle.

De Broglie explained, although not without difficulty, why the electron in the Bohr hydrogen atom orbited in stationary states. Only in the stationary states does the de Broglie wave of the electron resonate and not self-interfere. But de Broglie uses his original wave, a pure, monochromatic, and not the group of waves of somewhat differing wavelengths in this explanation. Some historians of science claim that de Broglie's ideas were incoherent. If they were, they were nevertheless ideas whose time had come, for Schrödinger's wave mechanics builds a new quantum dynamics on de Broglie's insight.

The two theories, of Heisenberg and Schrödinger, could hardly have appeared to be more different. The mathematical apparatus of wave mechanics was familiar to the community of physicists and rather like the differential equations of hydrodynamics. Heisenberg's matrices were entirely unfamiliar and highly user-unfriendly. Yet Schrödinger in 1926 proved the two theories empirically equivalent, at least as far as the stationary, or stable-orbit, values for dynamical variables were concerned. What the physicist nowadays tends to call 'elementary quantum mechanics' is a user-friendly blend of the two theories. What the contemporary mathematician calls quantum mechanics is an abstraction from both due to John von Neumann.

What provides the fundamental environment of the new physics in the period of the middle 1920s just before the development of quantum mechanics? First, there is the breakdown of continuous change in Nature, a breakdown which is explicit in Planck's law and in the Bohr theory of the hydrogen atom. Second, there is the duality of wave and particle in both radiation, in Einstein's explanation of the photoelectric effect, and, though it seemed even more dubious at the time, in matter, via de Broglie's positing of the wave aspect of material particles.

The former is a severe limitation of classical physics, but a limitation which is less disturbing than might appear because the old quantum theory was so obviously an unsatisfactory theory of fundamental physics. The second, wave–particle duality, is a paradox which persists in quantum mechanics proper, and which is one of the sources of the metaphysics which Schrödinger, Bohr and Einstein among others read into quantum theory.

What is quantum mechanics?

How should quantum mechanics look to the philosopher of physics? A simple, skeletal and unhistorical answer is this.

The formalism of quantum mechanics describes microphysical systems:

(1) by assigning to them a *state-vector* (or wave-function in Schrödinger's idiom), and state-vectors can be added or *superposed* to give new state-vectors; and

(2) by assigning to their dynamical variables (or observables, things like position, momentum etc.) certain corresponding *operators* on the set of state-vectors such that

(3) from a knowledge of a system's state-vector and the operator corresponding to any observable one can calculate the *probability* that a measurement of that observable on the system will have its result in any range we care to choose.

A state-vector is a mathematical representation of what is true of a quantum system, that is of its *state*. An operator is something that multiplies into a state-vector yielding a state-vector, usually different from, that is, not a numerical multiple of, the original state-vector. If it is a simple numerical multiple of the original state-vector then, for reasons we give later, the dynamical variable corresponding to the operator has that definite value in the state. No state-vector can, in fact, associate definite values with *all* the dynamical variables of the system.

Note how much more indirect is the interpretation of the formalism of quantum mechanics compared with that of classical mechanics. In the case of classical mechanics one handles the quantities of, for example, position and momentum, quantities which directly describe physical quantities possessed by or true of physical systems. In the quantum mechanical case the values of dynamical variables are associated with the effect of an operator on a quantum state description.

But the fundamental oddness of quantum mechanics in comparison with classical mechanics is that:

(4) certain pairs of operators A, B corresponding to dynamical variables *fail to commute*, that is AB is not equal to BA.

This means roughly that the two operators need not always have simultaneous determinate values. What it means exactly is a problem for the philosophy of quantum mechanics. To give noncommutativity the gloss we do is to introduce a controversial element of interpretation. For example, it seems to follow immediately that the values of dynamical variables cannot be determinate numerical values, for if they were the products of their values would be commutative (as multiplication in ordinary arithmetic is) and so would the operators that correspond to them.

But one can say this, pretty uncontroversially. Operators operate on state-vectors to produce (usually new) state-vectors, so that a state-vector can be operated on by a product of operators. Given a state-vector v and an operator A then Av (A operating on v, or A multiplies into v) will generally be a new state-vector. Another operator B may then operate on this state-vector to produce another new state-vector. We can think of $B(Av)$, or B operating on the outcome of A operating on v, as the operation of (BA) on v. Operators can be multiplied to form new operators and so it turns out that operators form an algebra. The odd thing about it is that it is generally a *noncommutative algebra*. Two operators A and B fail to commute if and only if there is a state-vector v such that

$$A(Bv) \neq B(Av)$$

so that

$$AB \neq BA.$$

The really interesting case is that of the pair of fundamental operators X, for position, and P_x, for momentum. For them, there is *no* state-vector v such that

$$XP_xv = P_xXv.$$

In fact, the difference $XP_x - P_xX$ is (oddly) a constant $i\hbar$, the square root of minus one times Planck's constant (divided by 2π). In other words, $[X, P_x]$, *the commutator* of X and P_x which is just shorthand for $(XP_x - P_xX)$, has the following effect on any state-vector v (in the domain of X and P_x and of XP_x and P_xX)

$$[X,P_x]v = (XP_x - P_xX)v = i\hbar v.$$

It multiplies the vector v by $i\hbar$.

Two important consequences then follow within the formalism of quantum mechanics.

(5) For the standard deviations (or spreads) δX and δP:

$$\delta X \delta P \geqslant \tfrac{1}{2}\hbar$$

which is the *uncertainty principle* or *indeterminacy relation* for X and P_x (what it means exactly is a problem in the philosophy of quantum mechanics), and

(6) if a state-vector vanishes except within a finite range of values of X, then it is unlocalizable in P_x and vice versa. This interesting result we may call the *complementarity theorem* for X and P. It has an arguable connection with the principle of complementarity of Niels Bohr. We refer to it in Chapter 3 and in Chapter 10.

Some comments

In expressing (1) to (6) as we have, we have begged important questions in the philosophy of quantum mechanics. The ensemble interpreter will object to associating a state-vector with an individual system rather than with an ensemble. Note too that the words 'probability' and 'measurement' appear in the fundamental rule (3) of the theory and so even the fundamental axioms present problems to the philosophy of quantum mechanics. Describe the formalism of quantum mechanics and you immediately encounter philosophical problems. Probability presents philosophical puzzles in itself. In classical physics there is no special problem about measurement. Measurement is simply an application of classical physics. But in quantum mechanics measurement is a fundamental and apparently ineliminable concept. It plays a privileged role, and this causes measurement to be a problem, perhaps *the fundamental* problem of the philosophy of quantum mechanics.

Classical and quantum mechanics compared

Classical and quantum mechanics clearly employ different notions of *state* and *observable*.

First, consider the quantum mechanical notion of *state*. The expression 'state-*vector*' for the mathematical description of the state of a quantum system is quite apt because we *superpose*, or add, quantum states. Take two quantum states, add them, and you have a third state, different from both original states, exactly as when in vector algebra you add two position-vectors using the parallelogram law to obtain a third vector. This is the famous *superposition principle*.

You can add quantum states just as you can add waves. You can also localize a quantum system as when you make a measurement of its position. These apparently contradictory facts derive in part from the superposition principle and are the essence of *wave–particle duality*.

They are reflected in two early rival interpretations of quantum mechanics, one due to Erwin Schrödinger and the other to Max Born.

What physical significance does the superposition principle have?

Consider two different states of a free electron, one in which its momentum is confined to a given narrow range, and the other in which its momentum is confined to a second narrow range which does not overlap with the first. Suppose you have an electron which is in a superposition of these two states. Now make a measurement of its momentum.

First, you will find that its momentum is in one of the two ranges. It will not be found to have 'the average' of the two momenta. Second, if you make a second measurement of momentum you will find that the momentum is in the same range. Repeat the measurement and you get the same result. The electron will be found to be permanently in its revealed state.

One way of describing this is to say that the original superposition is *projected* onto the revealed state by the first act of measurement. When two quantum states are superposed there is good reason to say that the new state is not a mixture of the two old states but a merging of them into a new state. Yet, in the case we imagined, measurement can recreate one of the original states. Which of the two original states is recreated on measurement is a matter of *chance*.

Contrast the case of classical mechanics. Of course you can take two particle states, or more accurately two state-descriptions

$$\langle x_1, p_1\rangle \text{ and } \langle x_2, p_2\rangle$$

of position and momentum, and 'add' them to obtain a new state, or rather state-description

$$\langle x_1 + x_2, p_1 + p_2\rangle$$

But this classical 'adding' *has no physical significance*. The new state-description does not describe a state which in any sense merges or even mixes the old states. Measurements of position and momentum will reveal the values $x_1 + x_2$ and $p_1 + p_2$. The quantal superposition of states is physically significant in that the new state is different from the states from which it was created.

So much for states. What about observables?

In quantum mechanics these are associated with operators. An operator maps a state-vector onto a state-vector. But that says almost nothing until you know what operators to associate with what observables. So where do the operators come from?

First there are position and momentum. Second there are observables and hence operators which can be derived from them.

Position and momentum are, in a sense, basic in quantum mechanics just as they are in classical mechanics. In fact, one can take the commutation relation for X and P_x to be the fundamental law of quantum mechanics.

One can then derive the other quantum-mechanical operators, like angular momentum, from the operators X and P_x using the corresponding classical expressions. Thus angular momentum in the z-direction L_z is just

$$L_z = XP_y - P_yX.$$

Here we use a kind of *generalized correspondence principle*. Here is a way in which the use of the quantum-mechanical formalism is dependent on classical mechanics. We need classical mechanics to tell us what the quantum-mechanical operators are.

Though position and momentum are fundamental in the formalism of quantum mechanics we shall as a matter of policy try to avoid them as much as possible in later chapters. We shall discuss what physics needs to be discussed using the example of quantum-mechanical *spin*. Spin has this great conceptual advantage over position and momentum: a quantum system can have only one of a finite number of possible different values for its components of spin. For example, in any given direction an electron can have only spin 'up' or spin 'down'.

Using spin we can simplify our treatment of quantum-mechanical puzzles like EPR, nonlocality in general, and certain features of quantum logic. Position and momentum both take a continuous range of possible values and this introduces awkward infinities (of the sort that Dirac's illegitimate δ-function was meant to handle).

Quantum heuristics

Now for some heuristics which take us from classical mechanics and the commutation relation

$$[X, P_x] = i\hbar$$

to the essential features of quantum mechanics.

We choose to represent the possible states of a quantum system as state-vectors which are functions of position x. How can we arrive at the operators X and P_x?

It seems reasonable that we represent X by the position vector x, in one dimension. What then of P_x?

Turning to the commutation relation, remembering that X and P_x are operators acting on functions, we have

$$[X, P_x] f = [XP_x - PX_x) f = i\hbar f$$

for any reasonable function f.

Substituting $P_x = -i\hbar d/dx$ we see that the commutation relation is satisfied. So we take $-i\hbar d/dx$ as the representation of the operator P_x.

Having got this far we can see where the Schrödinger equation comes from. A classical particle moving with momentum p in a field potential given by V has energy E given by

$$E = p^2/2m + V.$$

So the quantum-mechanical energy operator E is given by

$$E = (-h^2/2m)(d^2/dx^2) + V.$$

This operator acts on functions like f. The Schrödinger equation tells us that

$$(-h^2/2m)(d^2/dx^2) f + Vf = ef$$

where e ranges over the allowed stable and definite energy levels or energy eigenvalues and the fs which satisfy the Schrödinger equation are the energy eigenstates.

The ideas of eigenvalue and eigenstate apply generally to all observables. Thus a function f is an eigenstate of an observable **O** with eigenvalue o_n if and only if

$$\mathbf{O}f = o_n f$$

The idea that a particle can have definite value for a dynamical variable when and only when it is in an eigenstate of the corresponding operator is very important. Call it *the eigenvalue–eigenstate link*: a system has a definite value for a dynamical variable if and only if it is in an eigenstate of the corresponding operator, the definite value being the eigenvalue associated with the eigenstate. It is not an uncontroversial idea. After all it has to do with the interpretation of 'state'. But something can be said to make it plausible: if a measurement of **O** is made on a system in the eigenstate f then the probability that **O** yields o_n is one. What is controversial is the converse. Does a system have the **O**-value o_n *only* when it is in the eigenstate f?

To get moving we need examples of eigenstates and eigenvalues. The most attractive example to give would appear to be an eigenstate of momentum. Almost all basic textbooks of quantum mechanics give exactly this. Take the function

$$g = \exp[-ip_0 x/\hbar]$$

and (an exercise for the reader) show that

$$Pg = p_0 g.$$

Here p_0 is a constant with the dimensions of momentum and x is the position variable. Hence an electron whose state-vector is g will have determinate momentum p according to the eigenvalue–eigenstate link.

We will find almost immediately that there is something wrong with this example. What exactly?

To answer this question we must look at the way in which probability figures in the fundamental assumptions of quantum mechanics.

Suppose we know that a quantum system is in a state f which is *not* an eigenstate of **A**. Then what is the *expectation-value* of **A**? What, in other words, is the value of **A** we may expect? What is the average value of the results of measuring **A** on systems prepared to be in state f? The answer is given by thc fundamental algorithm of quantum mechanics which specifies the following procedure.

Take the state f and operate on it with **A,** getting $\mathbf{A}f$. Then integrate the product of $\mathbf{A}f$ with the complex conjugate of f, namely f^*. Finally divide this by the integral of $(f^*)f$. This gives you the expectation-value of **A** in the state f. One writes

$$\langle \mathbf{A} \rangle = \int f^*(\mathbf{A}f)\,\mathrm{d}x \,/\, \int f^* f\,\mathrm{d}x.$$

For this expression to make sense f must be square-integrable, otherwise the denominator will be undefined. Our momentum 'eigenstate' g isn't square-integrable and so cannot be allowed as a genuine state-vector. That is why we tend to avoid exact momentum and indeed exact position 'eigenstates' whenever we can. It turns out that although there are no exact, or point-position or momentum eigenstates, there are certainly eigenstates of 'position (momentum) restricted to a finite range'.

We can go further and simplify the expression for the expectation-value if we normalize state-vectors so that

$$\int f^* f\,\mathrm{d}x = 1,$$

so that

$$\langle \mathbf{A} \rangle = \int f^*\,(\mathbf{A}f)\,\mathrm{d}x.$$

It is easy to see that normalization involves no more than a multiplication by a constant and so cannot affect eigenvalues or expectation-values. So we let this simplified expression, with normalized eigenstates, be our version of the quantum-mechanical algorithm. Henceforth all eigenstates will be assumed to be normalized.

One last but important fact. In all the examples we consider in this book two different eigenstates of a given observable are *orthogonal*.

Two eigenstates are orthogonal when they are at 'right-angles', when in other words

$$\int f^* g \, dx = 0.$$

The integral of f^* times g over x behaves like a scalar product. The analogy is in fact very close as we shall see in Chapter 6. It turns out that the set of state-vectors has exactly the structure of a vector space, a space whose structure forms the main mathematical object studied by mathematical physicists.

The notation for expectation-values – $\langle \mathbf{A} \rangle$ for example – suggests a change in the way one writes state-vectors. Write the state-vector f as $|f\rangle$, and write the conjugate vector f^* as $\langle f|$. The first is called a 'ket' and the second a 'bra'. Then you can write the integral of f^* times $(\mathbf{A}f)$ over x as

$$\langle f | \mathbf{A} | f \rangle$$

the x and the integral sign being suppressed in the new notation, which we now adopt as our norm.

Often we drop all reference to fs, when there is no need to name them and write $|\rangle$ and $\langle|$ for a typical bra and ket. The 'bra' 'ket' device is known as Dirac notation and, although it seems to be no more than an elegant trick, it actually simplifies many of the expressions one runs into quantum mechanics.

Spin

Why must we bother with intrinsic electronic spin?

The answer comes from spectroscopic data and also from the classic experiment performed by Stern and Gerlach in 1922 which shows that there is an observable 'intrinsic spin' which, for an electron, has only two possible values in any given direction.

An oddity of the intrinsic spin of an electron in any given direction is that it can take only one of two possible values – 'up' or 'down'. A 'spinning electron' is quite unlike a spinning top. Quantum-mechanical spin is highly unclassical. In both classical and quantum mechanics we think of electrons as having an intrinsic mass. In quantum mechanics but not in classical mechanics they also have an *intrinsic* spin which is not to be thought of as a spinning motion. Yet an electron's spin has x-, y-, and z-components just like a spinning motion. So think of spin as an intrinsic property of an electron, a property which has the mathematical structure of an angular momentum.

Stern together with his assistant Gerlach sent a narrow beam of silver atoms in what was otherwise a vacuum through collimating slits and an inhomogeneous, that is varying, magnetic field produced by a strong electromagnet. When the beam had passed through the field it was made to strike a glass plate placed normally to its path. Stern and Gerlach found that the beam split into two sub-beams, one deflected 'up' by the field and the other 'down.' No atoms were left undeflected.

The explanation for this is given by the fact that each silver atom has half-integral spin, due to its outer electrons, which can have its component in any given direction either 'up' or 'down'. As each atom is charged, it behaves like a magnet which is deflected by the magnetic field. In principle, individual electrons exhibit exactly the same effect as silver atoms, the effect being called *space quantization.* Space is a honeycomb but that the spin direction of an electron or a silver atom is quantized. (In practice you can't do a Stern–Gerlach experiment with *electrons* because the magnetic field acts on their charges as well as their spin, blurring the effect.)

So we simply accept that in any chosen direction an electron has spin 'up' or 'down'. Therefore the state-vector which describes spin has only two components. The overall state-vector which describes an electron will be a product of first a nonspin state-vector and secondly the state-vector for spin. For any given direction z, the state-vector for the spin part must be a weighted sum of z-spin 'up' and z-spin 'down', the sum of the squares of the weights being 1.

To treat spin we appeal to the generalized principle of correspondence. A classical particle with coordinates x, y and z, and momenta P_x, P_y, P_z has its z-component of angular momentum l_z given by

$$l_z = xp_y - yp_x.$$

One can get the $x-$ and $y-$ components l_x and l_y from the equation by permuting x, y, and z. Substituting $P_x = -\mathrm{i}\hbar \mathrm{d}/\mathrm{d}x$, etc. one can then obtain the quantum-mechanical expressions for the angular momenta L_x, L_y, and L_z. Furthermore one can verify that the following commutation relation holds

$$[L_x, L_y] = \mathrm{i}\hbar L_z.$$

Again one can permute x, y, and z to get two more commutation relations.

How can we represent the operators S_x, S_y, S_z corresponding to the components of the observable 'spin'? The answer is: by means of the *Pauli spin matrices* which are:

$$o_x = \begin{pmatrix} 0 & 1 \\ 1 & 0 \end{pmatrix} \quad o_y = \begin{pmatrix} 0 & -\mathrm{i} \\ \mathrm{i} & 0 \end{pmatrix} \quad o_z = \begin{pmatrix} 1 & 0 \\ 0 & -1 \end{pmatrix}$$

If you substitute $S_x = \frac{1}{2}\hbar\ \sigma_x$ etc. then you can see that the commutation relations

$$[S_x, S_y] = i\hbar S_z, \text{ etc.}$$

for spin angular momentum are satisfied.

With the spin angular momentum operators represented by 2-by-2 matrices, the spin states are represented by column matrices with 2 elements.

What is the state-vector $|z, +\rangle$ for an electron having spin 'up' in the z-direction?

Since we must have $S_z|z, +\rangle = +\frac{1}{2}\hbar|z, +\rangle$ then, since the sum of the squares of the two components 'upper' for 'up' and 'lower' for 'down' must be unity, we have

$$|z, +\rangle = \begin{pmatrix} 1 \\ 0 \end{pmatrix}.$$

Similarly for spin 'down' you get

$$|z, -\rangle = \begin{pmatrix} 0 \\ 1 \end{pmatrix}.$$

A more interesting question is: what is the representation in terms of $|z, +\rangle$ and $|z, -\rangle$ for an electron whose spin is 'up' in the x-direction?

Let

$$|x, +\rangle = \begin{pmatrix} a \\ b \end{pmatrix}.$$

The term $|x, +\rangle$ is the state-vector for an electron whose spin is 'up' in the x-direction. We are asking how we should represent this state-vector in terms of a superposition of spin 'up' and spin 'down' in the z-direction.

As $S_x|x, +\rangle = +\frac{1}{2}\hbar|x, +\rangle$

$$\begin{pmatrix} 0 & 1 \\ 1 & 0 \end{pmatrix} \begin{pmatrix} a \\ b \end{pmatrix} = \begin{pmatrix} a \\ b \end{pmatrix}$$

and hence we can choose $a = b$. Using normalization we have $a = b = 1/\sqrt{2}$.

Using the Pauli spin matrices you can (and should) show that one can represent $|y, +\rangle$ in terms of spin 'up' and 'down' in the z-direction as

$$|y, +\rangle = \begin{pmatrix} 1/\sqrt{2} \\ i/\sqrt{2} \end{pmatrix}$$

Using the probability interpretation of the state-vector one can say that if an electron has spin 'up' in the z-direction the probability that it will be found to have spin 'up' in the y-direction after a measurement of spin in the y-direction will be $|1/\sqrt{2}|^2$, that is $\frac{1}{2}$. Similarly for spin 'down' in the y-direction and spin 'up' and spin 'down' in the x-direction.

Though conceptually odd, we shall find that spin is much more tractable mathematically than either position or momentum. Therefore, we shall handle EPR, nonlocality, and failures of distributivity in quantum logic using examples of spin and angular momentum.

The phenomenon of electronic spin is formally similar to the polarization of light. Light is electromagnetic radiation. Unlike sound, it is a transverse wave. The electric and magnetic fields can oscillate in a plane which is at right-angles to the direction in which the wave propagates. If the direction in which the electric field oscillates is fixed (it needn't be), the light is said to be *polarized,* and the plane at right-angles to the direction of propagation is called the plane of *polarization.* Light can be polarized in many different ways and by many devices, among which are with a polaroid, which simply screens off the light which is not polarized in the right direction, and with calcite crystals, which pass two beams of oppositely polarized light.

We can represent light which is thus polarized in a given direction (say at 30° to the vertical) into a superposition of light waves polarized vertically and horizontally (in our case with a weighting of $\sqrt{3}$ to 1: why?).

Formally, polarization in the vertical and horizontal directions is the same as spin 'up' and 'down', and the paradoxes which we illustrate with spin, like EPR and Bell nonlocality, may be handled equally well with polarized light. Of course, polarization itself is not a quantum mechanical phenomenon (except in the sense that Maxwell's equations are Schrödinger's equation for light). But the fact that polarized light comes in photons is quantal, and it is this which generates the paradoxes.

That is all the quantum mechanics we need. It will give us quantum-mechanical nonlocality. It will motivate wave–particle duality, the superposition principle and their mathematical descendant, quantum-logic.

3

Wave–particle duality

The simplest but deepest questions of the philosophy of quantum mechanics are these: Are quantum systems waves, or are they particles? Are they both or are they neither?

When light bounces an electron out of the surface of an alkali metal it behaves like a particle. Light carries momentum but also a punch. Just as there is no such thing as half a punch, so there is no such thing as half a quantum of light.

On the other hand, the shadow formed on a screen by an object has soft edges, edges which get softer as the object is moved away from the screen. Light exhibits the phenomena of diffraction and interference and so is wavelike.

This dual behaviour of light, and indeed of matter itself, is called *wave–particle duality*. The wave nature of a quantum system is reflected in wave mechanics which assigns a wave-function to the system. But the dual nature of light was noticed before Schrödinger developed the wave-mechanical formalism in 1926. In fact, it goes back at least as far as Einstein's explanation of the photoelectric effect in 1905.

Wave–particle duality is not something that arises within the quantum-mechanical formalism. It is something which is expressed within it, and is something which is best discussed in terms of the paradoxes that quantum experiments force upon physics.

In fact, the simplest and most immediately disturbing of the quantum-mechanical paradoxes arise out of wave–particle duality. We take the two-slit experiment and the delayed-choice experiment as examples.

The two-slit experiment

The two-slit experiment captures in one thought-experiment about half of what is puzzling in quantum mechanics and so it has been discussed by almost everyone who has written on the philosophy of quantum-mechanics. It is essentially nothing more than one of the interference phenomena noted by Thomas Young in his *Experiments and Calculations Relating to Physical Optics,* a lecture given on November 24th 1803.[1]

Interference phenomena of the type that Young describes are taken to demonstrate the need for a wave theory of light and Young's lecture is often seen as the start of the revival of the wave theory which came in the nineteenth century to replace the corpuscular view of light. Yet the quantum theory of light makes a paradox of the unparadoxical wave interpretation or the experiment. The two-slit experiment has a deceptive simplicity. But every interpretation of quantum mechanics must give an account of it which succeeds in avoiding paradox. Quantum logical interpretations of quantum mechanics are no exception as we shall see.

Here is the paradox. Imagine a beam of light, a stream of photons produced by some source or other. Imagine that the light from the source has been focussed to be parallel, moving towards a screen, which is the first screen of two. Imagine also that the light is *monochromatic,* that is, consists of light of exactly one wavelength. Of course, purely monochromatic light is an impossibility, as we noted in Chapter 2, but we can make our light as monochromatic as we please. And as a matter of fact, the experimental puzzle that makes the two-slit experiment famous is so gross that it will be exhibited by light which is only approximately monochromatic.

The first screen is called the *diaphragm* and is placed normally, that is, at right-angles, across the path of the beam. The diaphragm is meant to be a barrier to the photons. But two narrow, parallel slits have been cut in it, and the slits are just wide enough to allow only a dim beam of photons through. The slits can be open or closed. When they are closed they let no photons through. Even when they are open, most of the photons fail to negotiate the slits and are absorbed on or reflected from the diaphragm.

After negotiating the diaphragm those few photons that pass through either of the slits travel on to strike another screen placed some distance behind the diaphragm but lying in a plane parallel to it. Imagine the distance between the diaphragm and the second screen as being large compared with the distance between the slits.

What do we observe? The first important point is that each photon strikes the second screen at a definite location. Each photon behaves like a particle. It is *not* spread out over the whole screen. The screen may be a photographic plate and the pattern of photon hits may be recorded as a pattern of blackening on the photographic emulsion. One can think of the second screen as being a position measuring apparatus which produces as a result a definite location for each photon that hits it, so that at the screen each photon behaves like a particle. At the screen, light exhibits corpuscular behaviour. Therefore photons are particles.

Now consider what happens between the front face of the diaphragm – the first screen – and the screen on which the photons are localized. There are three cases to consider. We have two slits, and we can have either one or the other open, or both open together. Of course, if they are both closed nothing happens.

First, suppose that *slit 1* is open and *slit 2* is closed. Naturally the intensity of the photon hits on the screen is less than when both the slits are open. The frequency of their hits is halved overall. The interesting thing is their distribution over the screen. This is characteristic of the diffraction pattern for waves passing through the single slit. Each photon appears as a particle. The region on the screen of most likely impact is behind the open slit but there is also a considerable spread of impacts either side of this position.

In the second case, if *slit 1* is now closed and *slit 2* now opened one obtains an exactly similar result except that the whole pattern of impacts on the screen is now shifted so that its center is directly behind *slit 2*.

The really interesting case is the third, in which both slits are open. First of all there is a natural increase, in fact a doubling, in the mean number of hits per unit time on the screen as a whole. But the distribution of those impacts is quite unexpected. The new distribution is *nothing like* the sum of the two single slit patterns. There is a maximum, a region of maximum frequency of hits near the position on the screen opposite the center of the gaps between the two slits. But there are plenty of other maxima as well. Even more interesting, there are minima, regions almost free from photon hits in between the maxima of the two-slit pattern, *in positions where the sum of the two separate single-slit patterns is not zero.* There are places where the two slits produce fewer hits than each slit separately. These are where *destructive interference* occurs. The locations of the local maxima – the positions on the screen where the photons are more likely to hit than on immediately neighboring positions – are positions where *constructive interference* takes place.

What is it that is so paradoxical?

One can explain the behaviour of the photons on the second screen by assuming them to be particles since there they appear to be localized. One can explain their diffraction and (in the two-slit case) their *self-interference* by assuming them to be waves. The striking thing is that one cannot give an explanation of the whole experiment by consistently assuming them to be one or the other.

One can express the problem in the following way.

If the photons are particles it should not matter whether or not the slit which they do not go through is open or closed; the two-slit pattern should have the form of the sum of the two single-slit patterns.

But if the photons are waves one cannot explain how they come to be localized when they hit the screen. Assuming them to be waves (or *wave-packets*) one says that at the screen there is a *collapse of the wave-packet* (since a photon is not a single monochromatic wave, but is rather a collection, or superposition, of waves). We meet this picturesque and perhaps rather unfortunate phrase again when we discuss measurement. But how the collapse, if it is meant to be a physical effect, can occur is quite unexplained.

The classical explanation of the two-slit experiment, that is of Young's experiment, is of course unparadoxical. It *simply fails to note the particle behaviour on the screen* and so it avoids paradox by not even considering, let alone admitting, the particle nature of light. But the quantum account is paradoxical in that it assigns contradictory properties to the photons.

Some points to note.

First, the experiment can be performed with electrons as well as photons, though in the case of the electrons the diaphragm is usually a crystal lattice and the slits, many rather than two, are the spaces between atoms in the lattice. The classic experiment with electrons was performed by Davisson and Germer in 1927, and is mentioned in every textbook on quantum mechanics.

Second, nothing in our discussion of the experiment depends upon any mutual interaction of the separate photons that go through the slits. Diffraction experiments with very low intensity light, so low that only a single photon is diffracted at any time, give exactly the same results as higher intensity diffraction experiments. The earliest experiments showing this were performed by Sir Geoffrey Ingram Taylor[2] in 1909. The light source that Taylor used was so low in intensity that one of his experiments took three months to produce a blackening of his photographic plate. Yet he found that the diffraction pattern produced by light which passed a needle was independent of the intensity of the light. Therefore, the two-slit experiment both as a thought-experiment and in its practical realizations embodies a puzzle about the *self-interference* of photons.

Third, any attempt to find out which of the two slits each photon goes through destroys the characteristic two-slit interference pattern.

One can set up detectors to register the slit that a given photon passes through but as these give better and better knowledge of the path of each photon so the pattern goes over from the undisturbed two-slit pattern to a pattern having the form of the sum of the two single-slit patterns. In other words, if you know which slit the photon went through, then the photon's self-interference is interfered with.

So what is the outcome?

We seem to have to say that light is both corpuscular and wavelike – we have to say that the photon, a particle, goes through *both* slits when they are open, that a photon interferes with itself, and yet behaves as a particle when it hits the screen.

Wheeler's delayed-choice experiment

The delayed-choice experiment is a version of the beam-splitter thought-experiment which Einstein and Bohr argued over in the course of their protracted dialogue,[3] an experiment which is very similar to the two-slit experiment and which, both as a thought-experiment and as a real phenomenon, has been revived by John Archibald Wheeler.

In the beam-splitter experiment one has a beam of photons which is divided into two sub-beams – just as it is in the two-slit experiment. This time the divider is not a screen with two slits but a half-silvered mirror set at 45° to the direction of the beam. One sub-beam passes through the half-silvered mirror while the other is reflected through a right-angle. Two further fully silvered mirrors then reflect the two sub-beams through right-angles so that the sub-beams make a rectangle and pass one another at 90° at the corner opposite to that of the half-silvered mirror.

Photon counters are then placed to detect the two sub-beams after they have crossed one another. The result is the one you would expect on a particle picture. There is no interference between the beams since they are travelling perpendicularly to one another. Arrivals at the two counters are independent of one another, exactly as if each photon were a particle which took only one of the two possible paths through the beam-splitter. So we can say in this case, without fear of paradox, that each photon went through just one path through the beam-splitter. In fact, if the photon were to take both paths, it would be hard to understand why it should appear to have taken just one of the two paths, why, that is, it is detected at A (say) rather than at both A and B.

There are then two interesting things one can do.

Interesting thing number one. Place a second half-silvered mirror, inclined exactly as the first, at the point where the two sub-beams pass one another. Now the two paths taken by the two sub-beams can be varied slightly so that they can be made to interfere in such a way that either of the two counters fails to detect any photons, in which case destructive interference must be taking place at that counter. This version of the experiment cries out for a wave picture of photons. Since we can arrange to have either destructive or constructive interference by varying the difference (or otherwise) in the lengths of the two possible paths, we have to say that each photon goes via both sub-beams in the apparatus. Of course, just as in the case of the two-slit experiment, the delayed-choice experiment can be repeated with very low intensity beams having only one photon at a time in the beam splitter and the result will be the same. So far we have the two-slit experiment in a somewhat different form.

Interesting thing number two. First, note that we can make the two paths through the beam-splitter as long as we please. Second, note that we can choose to insert or not to insert the second half-silvered mirror *after* the photon has begun to traverse the beam-splitter. We can *delay making our choice* as to whether the photon must take just one path through the beam splitter or both paths, which is why the delayed-choice experiment is so called.

The beam-splitter experiment presents wave–particle duality in a particularly stark form. It can be explained without a detailed treatment of interference. It brings out the fact that how the phenomenon appears to us, whether we say that the photon goes through both or only one of the beams, depends on decisions we make, something which is fundamental to later versions of the Copenhagen interpretation.

Wave versus particle

The quantum-mechanical wave–particle duality which the two-slit experiment and the delayed-choice experiment illustrate applies equally well to matter as to light. In the case of light, wave and particle theories had competed at least from the time of Newton's *Opticks*. That light may be both (or neither) wave or particle is a characteristically twentieth-century idea, as is the idea that what goes for light also goes for matter, and the idea that the dualism of light and matter is merely superficial.

As a philosopher of physics, one should ask what the essential difference between wave and particle theories of light is.

In any particle theory, light or matter is viewed as a *localized substance* which is *transferred* from a source to its target. The emphasis here should be on 'transference' rather than on 'substance'. A wave theory, on the other hand, views light or matter as an oscillatory motion which is not itself transmitted from source to target. In a wave something more abstract than a quantity of substance is propagated. What is moved from source to target is perhaps a disturbance or perhaps energy. Therefore a particle theory can incorporate an oscillatory motion of the transmitted substance without becoming a wave theory. In fact Newton's corpuscular theory of light does just this.

In the *Opticks* Newton developed his corpuscular view of light in such a way as to take account of diffraction and interference. His corpuscles moved with 'fits and starts' whose periods corresponded to the wavelength of light as that appeared in the wave theory. In Newton's own lifetime a wave theory of light was proposed by Christiaan Huyghens in his *Traite de la Lumiere* of 1690. But Newton's prestige was such that it was not until after the experiments of Thomas Young and the theoretical work of Fresnel and Arago in the nineteenth century that the wave theory displaced the corpuscular hypothesis.

Wave–particle duality – the apparent synthesis of both types of theory – seems early on to have been a consequence of the quantum hypothesis. It is implicit in Planck's law since Planck permitted the exchange of energy between matter and field only in discrete quanta. But this was a feature of his theory of black-body radiation that he disliked and which most contemporary physicists refused to accept. Wave–particle duality is also an explicit and intrinsic feature of Einstein's explanation of the photoelectric effect. Like Newton's corpuscles Einstein's free light quanta retain their identities in transit through space after which they can bounce electrons out of metal surfaces.

To the physicists of the time, Einstein's free light quantum was even more suspect than was Planck's quantum hypothesis. A fuller history of wave–particle duality in early twentieth-century physics would probably have to lay greater stress on X-rays than on the Einstein's free light quantum.[4] X-rays were discovered in 1895 by Röntgen but were generally thought to be high-frequency electromagnetic waves until 1912 or when they were shown to exhibit interference effects. Before that the dominant theory took them to be electromagnetic impulses. In fact G. I. Taylor's experiment of 1909 – which we mentioned in connection with the two-slit experiment – was performed to test J. J. Thomson's theory that X-rays, and possibly even light, were instances of electromagnetic waves propagating in a coarse-grained aether whose effect was to deliver electromagnetic impulses.

Niels Bohr was strongly opposed to the free light quantum, which might seem a little odd. It was only with the experiments of R. A. Millikan in 1916 that the photoelectric effect was fully accepted as a physical phenomenon by the community of physicists. It was not until the 1920s and the discovery of and explanation of the Compton effect, in which electrons exposed to radiation behave as if they are rebounding from collisions with particles of light, that the free light quantum idea became orthodox. Even the philosopher of physics should not forget that the history of physics is always more messy than our later reconstructions of it make it appear. The very word 'photon' was invented as late as 1926 and not by Einstein, but by the American physical chemist G. N. Lewis, who had some unusual ideas of his own about them which were not at all like Einstein's.

But at roughly the same time as this success for Einstein's corpuscularianism, namely in the early 1920s, there was another push away from corpuscularism. In 1924 Louis de Broglie revived the wave hypothesis and extended it from light to matter as a whole. De Broglie's theory, as he presented it in his doctoral dissertation, is strange and, one must say, blatantly unsatisfactory. Yet it expressed an idea whose time had come. For out of the chaos, the chaos of the wave–particle duality of light and the chaos of de Broglie's ideas, emerged in 1925–6 Schrödinger's wave-mechanics.

Schrödinger's wave ontology

Schrödinger supposed that the solutions of his equation represented *smeared out individual quantum systems.* He imagined reality to be wavelike. In his view, the fundamental parameter describing objects in the world was to be a frequency associated with the object – the frequency of the object's wave-function – and not its energy which is the fundamental parameter in a particle description. Schrödinger's ontology came down firmly on the side of waves.

Something of the wave ontology persists in the typical quantum chemistry textbook where electrons in atoms and molecules are thought of as electron clouds, a source of puzzles to intelligent students of chemistry (who recall the particle aspects of electrons) down to the present day. So the wave ontology, like most inadequate ideas, is useful for some purposes and offers an unexceptionable way of looking at certain problems. Unexceptionable that is, apart from being false. There are several standard objections to the wave ontology. Some of these are more serious than others.

What is wrong with the wave ontology?

First, there is the apparently merely technical problem that wave-functions are functions on (whose domain is) configuration space and not ordinary three-dimensional space. This shows up as a difficulty as soon as you consider the wave-function of more than one particle. For example, when you have the wave-function of an N-particle system you will find that the domain of its wave-function is a space of $3N$ dimensions, three for each particle. This $3N$-dimensional wave-function evolves smoothly according to the corresponding multidimensional Schrödinger equation. But each system is in real three-dimensional space. So one cannot identify the wave-function (or the square of the absolute value of the wave-function) with any physical system.

If you project out the wave-function for each particle by integrating over all the other variables, then you do have a wave-function for each particle which is in its configuration space, and this three-dimensional space can be thought of as real physical space. But then there are new difficulties. The particles behave nonlocally. Since the overall wave-function is well behaved in the $3N$-dimensional configuration space the behaviour of each particle's projected out wave-function depends on the wave-functions for all the other particles in a way that looks like action at a distance.

But the real trouble with Schrödinger's ontology is more a matter of physics than of mathematics. Thus, as a second objection, imagine a half-silvered mirror and let a photon strike it at 45° with an even chance of being reflected and an even chance of being transmitted. The interaction with the mirror changes the wave-packet from being an approximately pure plane wave moving in a fixed direction to being a superposition of two approximate plane waves, one moving in the original direction and the other moving at right-angles to it, both waves naturally moving away from the mirror. After a while the two wave-packets will have moved a large distance apart. Yet if a measurement were made to decide which of the two paths, reflected or transmitted, the photon really took, the answer would be the one or the other. You never observe half a photon. So you must say that the measurement caused the half a photon that was not observed to collapse into the half that was observed.

A third objection, also a matter of physics, is that Schrödinger's wave interpretation cannot explain Planck's law, that energy is exchanged between matter and radiation field in quanta whose size is determined only by the frequency of the radiation. This was the objection that Heisenberg pressed on Schrödinger when he heard Schrödinger lecture in Munich in July 1926 and also when Schrödinger visited Bohr and Heisenberg in Copenhagen in September of that year. Planck's law rules out a continuous exchange of energy between matter and field and demands *quantum jumps* which were inconsistent with Schrödinger's continuous wave ontology. In Heisenberg's version of the encounter in Copenhagen Schrödinger:

burst out almost desperately, "If one has to go on with these damned quantum jumps, then I'm sorry that I ever started to work on atomic theory."[5]

Finally, if an individual electron is supposed to be 'smeared out', it is sensible to ask 'how smeared out?', assuming that the technical objections have been overcome and we can think of an electron as its Schrödinger wave in real physical space. From the complementarity theorem we know that the wave is spread out through the whole universe if the momentum of the electron is restricted to a finite range, as we assume it is. Of course, it is true that the wave has most of its 'intensity' in a small range of physical space. But if the wave is what there is, it will be everywhere where its intensity is nonzero. Of course, this need not be a killing objection to Schrödinger though it does make his ontology implausible. The Schrödinger wave would not in any case seem like a good basis for physical ontology. In another form, however, this objection is also damaging to quantum logical attempts to reconcile quantum mechanics with realism, as we shall see in Chapter 10.

Born's particle ontology

If the wave ontology seems natural for light, at least to the classical mind, then it seems anything but natural for matter. The common resolution of wave–particle has been its collapse in favour of the particle. One finds a preference for the particle view in a whole variety of quantum-mechanical realisms, from the naive Popperian view that quantum systems are particles, to the sophisticated quantum logical view that wave–particle duality and the indeterminateness of the quantum world can be smoothed away with quantum logic.

If Schrödinger was the engineer who applied his knowledge of differential equations to atomic physics then Max Born was the mathematically sophisticated working physicist familiar with the physics of atomic collisions.

When a quantum system is scattered from a nucleus you observe it after scattering as a particle. Born's original interpretation of the wave-function viewed the quantum system as a particle travelling in a *guiding field* described by the wave-function. Schrödinger collapses wave–particle duality in favour of waves. Born, in his original probabilistic interpretation, collapses it in favour of particles. As he put it, the waves of wave mechanics:

are present only to show the corpuscular light quanta the way . . . the guiding field, represented by the scalar function ψ of the coordinates of all the particles involved and the time, propagates in accordance with Schrödinger's differential equation. Momentum and energy, however, are transferred in the same way as if the corpuscles are determined only to the extent that the laws of energy and momentum restrict them; otherwise, only a probability for a certain path is found, determined by the value of the ψ function.[6]

But the guiding field and how it guides are unexplained. Einstein, whenever he used the phrase in discussion, seems always to have used it humorously and sarcastically. The guiding field is not supposed to be a physical entity. If it is not, then how can we explain interference and diffraction? The trouble with Born's particle ontology is exactly the trouble with all particle pictures, the fact that they handle only one half of wave-particle duality, leaving the other half a mystery.

The same can of course be said of Schrödinger's wave ontology.

In the case of the two-slit experiment Schrödinger cannot explain particle behaviour at the screen and Born cannot explain the wave behaviour everywhere else without the mystery of the guiding field.

The failure of both these simple alternatives due to Schrödinger and Born paved the way for the Copenhagen interpretation – or the 'reigning doctrine' as Schrödinger later put it – which, in Schrödinger's words, 'rescues itself or us by having recourse to epistemology'.[7] The Copenhagen interpretation is more than a mere interpretation of quantum mechanics. It is a philosophy of physics in the broadest sense, a view of what physics can and cannot be expected to do. For that reason, it is where the *philosophy* of quantum mechanics can be said to start.

4

The Copenhagen interpretation (I)

Of all the many rival interpretations of quantum mechanics, none is more important or influential than the Copenhagen interpretation. Unfortunately, there is no one Copenhagen interpretation of quantum mechanics. If one uses the expression, and everyone does, one should not think of the Copenhagen interpretation as a single consistent interpretation of the theory.

Philosophers of physics tend to use the expression 'the Copenhagen interpretation of quantum mechanics' as a label covering a variety of different interpretations all of which either originate with Niels Bohr or were invented by him, his colleagues and guests at the Institute for Theoretical Physics in Copenhagen beginning in the middle 1920s. Perhaps one should say that the Copenhagen interpretation in this diffuse sense goes back farther than this, back even before the rise of the new quantum mechanics.

Heisenberg tells the story of a boat trip he, Bohr, and some friends took from Copenhagen to the island of Fyn around the time when the then new quantum mechanics was being developed. Bohr, 'full of the new interpretation of quantum theory', talked of 'the difficulties of language, of the limitations of all our means of expressing ourselves' and of how these limitations had been expressed 'in the foundation of atomic theory in a mathematically lucid way'. 'Finally,' Heisenberg ends the story, 'one of the friends remarked drily, "But, Niels, this is not really new, you said exactly the same ten years ago".'[1]

What we call the Copenhagen interpretation changed with time. There is the Copenhagen interpretation of the younger Bohr writing when quantum mechanics was fresh. There is the Copenhagen interpretation of the elder Bohr writing after the EPR paper of 1935. There are the applications of the principle of complementarity that Bohr makes to the problems of the nature of life and mind. There are the brief versions of the Copenhagen interpretation one finds in opening chapters of the standard textbooks, books written by authors who are quite justifiably anxious to get off the philosophical material and on with the physics. There are Heisenberg's versions of the Copenhagen interpretation. There are the works of Bohr's disciples like Rosenfeld and Bohr's expositors, like von Weizsacker.[2]

One can certainly say that the Copenhagen interpretation is quite different from the wave and particle interpretations of Schrödinger and Born. The Copenhagenists, and Bohr especially, wanted to draw from quantum mechanics not merely an interpretation of quantum theory but also general conclusions for epistemology and the philosophy of science, conclusions that would force us to recognize the limitations that either the microphysical world, or the structure of explanation in microphysics, (or perhaps both) impose on theoretical physics as an attempt to describe and provide a picture of the world. The Copenhagen interpretation is not about whether or not quantum systems are waves or particles or both or neither. It is a philosophy of physics, a philosophy which confronts the limitations on the *representations* that physics may employ, a philosophy of physics-as-a-cognitive-activity. It is deep. That is, it is extraordinarily unclear as to what the Copenhagenist philosophy of physics asserts.

The Copenhagen interpretation as physics and as philosophy is, among other things, Bohr's life's work. As Heisenberg put it

> Bohr was primarily a philosopher, not a physicist, but he understood that natural philosophy in our day and age carries weight only if its every detail can be subjected to the inexorable test of experiment.[3]

Like Wittgenstein's, Bohr's philosophizing (though performed with a rather better humor) was an activity of daily self-renewal. There is a view of the Copenhagen interpretation, sometimes put about by Popperians, that it was an iron dogma whose adherents would tolerate no deviation from the Copenhagen line. If this was true of some of Bohr's followers, it was not true of Bohr himself. Bohr constantly referred to the *purely formal* or *symbolic* nature of quantum mechanics, and indeed of his own solution in 1913 to the problem of the hydrogen atom, by which he meant to stress that quantum mechanics and the old quantum theory were adequate for the purposes of calculation but not as fundamental physical theory. Devastating philosophical consequences follow from the proposal, generalizing our experience with quantum mechanics, that *all* microphysical theories must suffer from this defect, the defect that we seem to be forced into producing physics theories which are adequate instruments for prediction but which are also inadequate vehicles for a fully determinate picture of the microphysical world.

Several themes run through the various accounts of the Copenhagen interpretation. One is a matter of methodology, an emphasis, to my mind a thoroughly praise-worthy emphasis, on thought-experiments. Like Einstein, Bohr had the physicist's preference for the simple physical picture over the mathematical formalism. One of Bohr's main philosophical concerns was with the relation between abstract theory and concrete experiment. He always sought to bring out 'the physics'. A second and corollary theme, more prominent in Bohr's writings than in those of the others, is a playing down of the formalism of quantum mechanics. To quote Heisenberg

> mathematical clarity had in itself no virtue for Bohr. He feared that the formal mathematical structure would obscure the physical core of the problem, and in any case, he was convinced that a complete physical explanation should absolutely precede the mathematical formation.[4]

It is difficult to account for the divergences within the Copenhagen school, of which we take Bohr and Heisenberg to be the principal representatives, without mentioning the differences between the two men. Heisenberg was the mathematical physicist whose chief interest lay in developing and applying the formalism of quantum theory. His main contribution to the Copenhagen interpretation lay in deriving the indeterminacy relations which he saw as the quantum-mechanical result of the greatest epistemological significance. In Heisenberg's view, the indeterminacy relations made meaningless all statements asserting a quantum system's simultaneous position and momentum to within ranges narrower than the indeterminacy relations allowed.

Of the two, Bohr had the more darkly metaphysical mind. He was the philosopher who, as Heisenberg put it, operated in the mode of theoretical physicist. He was preoccupied with that central question of any philosophy of physics: to what extent may the physical world be described? The thesis from which Bohr's answer springs is fundamental in his interpretation of quantum mechanics. We identify it – since like any commentator on Bohr's work we are and must be in the business of *organizing* as much as criticizing his thought – as the thesis of the priority of classical concepts. It arises from the following reflection. In whatever way we choose to describe quantum systems we must do so using classical concepts. Why? Because all the data by means of which we test our theories must be expressed in the language of classical physics. But why so? Because the language of classical physics is ideally suited to describing the macrophysical world, the world of medium-sized objects, in which we learn and constantly test natural language. The species has employed classical concepts (of position, speed and so on) for so many generations that they are now an intrinsic part of our world view. However far we travel from the world of everyday reality we are stuck with classical concepts.

This idea, that we are stuck with classical concepts, hangs together nicely with the correspondence principle, that the values of observables generated by a quantum explanation of a phenomenon must go over to the classical values in the limit of large quantum numbers. For example, as the energy levels of the bound electron in the hydrogen atom increase, as the electron orbits farther and farther from the nucleus, the gap between the allowed energy levels decreases so that it appears that the electron has a continuous range of allowed energy levels, which in fact it does not.

Following on from the correspondence principle and the fact of quantization, the fact of the nonzero quantum of action or the *quantum postulate,* comes the *principle of complementarity* which, in one of its many forms, is the idea that the classical concepts with which we are stuck have only a limited applicability to the quantum world.

This much is a sketch, a very brief sketch, of one way of looking at parts of Bohr's interpretation of quantum mechanics. It is a subject to which we return following our examination of the lighter metaphysics of the youthful Heisenberg.

Heisenberg and the γ-ray microscope

Heisenberg reports[5] that as a philosophical youth he was unimpressed by the drawings in a popular science text of atoms with hooks and eyes. The hooks and eyes held the atoms together in chemical compounds. Heisenberg naturally felt that understanding large physical structures involved accounting for them in terms of the behaviour of their parts, and that ultimate understanding involved understanding the ultimate parts. But he came to feel that such understanding might be possible only through a mathematical and not an iconic or graphic description (of the 'hooks and eyes' type).

The difficulties encountered by the Bohr–Sommerfeld theory of the atom in the early 1920s reinforced these speculations. The orbits that atomic electrons were supposed to take were unobservable and hence, to a mind of positivist leanings, philosophically suspect. More seriously, the old quantum theory had disconnected electronic orbital frequencies from the observed radiation emitted by the atom and so there was no good *physical* reason for demanding that electrons have orbits.

Heisenberg's great contribution to the interpretation of quantum mechanics was the indeterminacy relations. In his classic 1927 paper 'The physical content of quantum kinematics and mechanics'[6] he introduces the γ-ray thought-experiment to illustrate them. Yet his paper does much more than this. It is essentially a critique of the concepts of position and momentum as applied to the microphysical world, a critique in which the γ-ray microscope, along with other thought-experiments, appears only in passing.

What do we mean by 'position'? According to Heisenberg

> [w]hen one wants to be clear about what is to be understood by the words 'position of the object', for example of the electron (relative to a given frame of reference), then one must specify definite experiments with whose help one plans to measure the 'position of the electron': otherwise this word has no meaning.[7]

So let us try to specify a definite experiment to determine the position of an electron. We shine light onto it, and collect in some optical detecting gear (a microscope) some photons (or optimally, exactly one photon) which bounce(s) off it. Let the wavelength of the light – it is of course wave and/or particle – be q. The best we can hope for, according to Heisenberg, is to locate the electron to within a distance of the order of the wavelength q. This much we know from elementary classical wave optics. However, according to the laws of the Compton effect the electron will recoil with a momentum of the order p where

$$pq \sim h.$$

Therefore, since experiment cannot reveal it, the electron has no determinate location if its momentum is well-defined, and vice versa. Heisenberg then argues that this result, or one which is formally very like it, also follows from the commutation relation $[p, q] = -i\hbar$, and he demonstrates what is essentially the standard result that

$$\delta p \, \delta q = \tfrac{1}{2}\hbar$$

for the case of a *Gaussian* wave-packet.

In his γ-ray microscope example one must think of Heisenberg as employing a *reductio ad absurdum* argument, one of whose premisses is the operationist assumption that if no experiment is available to measure a physical magnitude then it is meaningless to assert that it has a value at all. Heisenberg first assumes that an observed electron has determinate position and momentum, and second that these are disturbed by its interaction with the photon used to see it. Therefore its position and its momentum cannot both be observed, and its position and momentum cannot both be determinate, because it would be meaningless to assert them both.

Interestingly, Heisenberg's analysis of the γ-ray microscope is incomplete, as Bohr noticed, and this fact explains the odd 'note added in proof' which appears at the end of Heisenberg's paper and in which he confesses to several errors in his treatment of the various thought-experiments. Heisenberg had failed to take into account the allowed spread in the momentum of the photon which recoils from the electron. The velocity of the recoiling photon may have any direction within the cone subtended by the lens of the γ-ray microscope and this, given the law of conservation of momentum applied to the electron/photon collision, is the real source of the indeterminacy in the momentum of the observed electron. The position of the collision is indeterminate because the image in the microscope produced by the photon is unsharp, for reasons provided by classical optics.

Heisenberg's conclusion from his thought-experiment is that quantum mechanics requires a revision of the concepts of position and momentum as applied to a quantum system. He also argues, as an aside, that quantum mechanics requires no revision in the *geometry* of space-time at small distances since the position and momentum of a physical object of sufficiently large mass may be measured with arbitrary precision. However the *joint application* of position and momentum to a small system such as an electron, as in the notion of an electron's 'path' or 'trajectory', has no definable meaning.

Of course Heisenberg requires an implicit assumption that *all* attempts to measure joint positions and momenta will fail just as they do in the case of the γ-ray microscope. That they do fail in the whole variety of imaginable cases is part of the case Heisenberg made in his Chicago Lectures of 1930, *The Physical Principles of Quantum Theory*.[8] He also seems to make the operationist assumption that 'joint position and momentum' have *meaning* only when there is an experimental procedure to determine both simultaneously. However, his argument that the geometry of space-time need not be revised is not operationist, but is dependent on the theoretical analysis of indeterminacy that his paper contains.

One can represent Heisenberg, at least at this time, as having a superficially positivist or operationist tendency. Less superficially he was what many philosophical physicists are, philosophers who base a metaphysics on the results of theoretical physics. He thought that the idea of electron orbits is incoherent and all talk of them meaningless because they are unobservable *according to the theory*. It was from Einstein that Heisenberg had learned that it is theory which decides what is and is not observable,[9] but he represented this idea in empiricist, or operationist guise, and from a great physicist one should not expect too much consistency of the type that philosophers value.

A more serious inconsistency is that in other places Heisenberg writes as if it is only our *knowledge* of the world that is indeterminate, not the concepts which we take to apply to physical systems big and small:

> The uncertainty principle refers to the degree of indeterminateness in the possible present knowledge of the simultaneous values of the various quantities with which quantum theory deals . . .[10]

If our knowledge falls short then presumably there is something more to be known, but in that case position and momentum are presumably simultaneously determinate, and it must be an incompleteness in quantum theory that leads to our incomplete knowledge of their determinate values. Uncertainty as limited knowledge fits in with a *disturbance* view of the measuring process, a view that Heisenberg sometimes put forward. In the γ-ray microscope thought-experiment, the electron is disturbed by its interaction with the photon. The disturbance is not entirely predictable, and this is the origin of the 'uncertainty'.

Heisenberg's formulation of the thought-experiment lends support to this interpretation. We argued that his conclusion was a *reductio ad absurdum*: the electron does not have simultaneous determinate position and momentum. Therefore any disturbance – and there may be no disturbance – cannot be the source of the uncertainty. Bohr was critical of Heisenberg's lapses into talk of the unavoidable disturbance during measurement as the origin of the indeterminacy.

In Bohr's view, wave–particle duality, the indeterminacy relations, and the completeness of quantum mechanics were all closely interrelated. Their connections are central to the Copenhagen interpretation and lead us on to the principle of complementarity.

The metaphysics of complementarity

Niels Bohr went to great lengths to refine and then to clarify his thoughts on quantum mechanics but, sad to say, his writings do not appear to have benefitted from these efforts. It is very difficult to say how his ideas – of the principle of complementarity, the priority of classical concepts, the quantum postulate and the methodological rule of the correspondence principle – hang together. We shall view the principle of complementarity as deriving from more fundamental principles, those of the priority of classical concepts and the quantum postulate.

First, the idea of the priority of classical concepts.

There is a sense, common to both the old quantum theory and to quantum mechanics, in which classical concepts are prior to the *application* of quantum theory. Using the old quantum theory we quantize a classical description. Classical physics is restricted so that stationary periodic motions are allowed only a discrete set of values dependent upon Planck's constant. In using quantum mechanics we derive our operators – especially the energy, or Hamiltonian, operator – from their prior classical expressions. This was how in Chapter 2 we got to Schrödinger's equation from the commutation relation for X and P. But Bohr insists on a deeper dependency of the quantal on the classical. Bohr thought that classical concepts were *indispensable* to quantum physics:

> however far the phenomena transcend the scope of classical physical explanation, the account of all evidence must be expressed in classical terms. The argument is simply that by the word 'experiment' we refer to a situation where we can tell others what we have done and what we have learned and that, therefore, the account of the experimental arrangement and of the results of observations must be expressed in unambiguous language with suitable application of the terminology of classical physics.[11]

The idea of the indispensability of classical concepts to any future microphysics is part of the justification of Bohr's remarkably successful methodological rule, the correspondence principle: when you are looking for a quantum explanation of a phenomenon, take care to make the limiting case of large quantum numbers converge to the classical values. But this success hardly forces classical concepts on us forever. Both Schrödinger and Einstein, for example, expected classical concepts to be superseded in a satisfactory new physics.

It is clear that Bohr exaggerates our reliance on classical concepts. What we need to communicate experimental results is a language which is both publicly accessible and which, since the macroscopic reality we are describing is determinate, is adequately determinate. But these demands do not imply that we cannot escape from the language of classical physics which, incidentally, is not the language of common sense but the highly un-common-sensical theoretical language of a changing science.

As a corollary to the priority of classical concepts, it follows that the classical concepts of position and momentum must necessarily figure in a quantum-mechanical description of quantum systems. Does this mean that they apply to quantum systems in the way that they apply to classical particles? Not according to Bohr. Position and momentum have a *limited applicability* to quantum systems, a limitation described quantitatively by the uncertainty principle, or as we say, 'indeterminacy relations'.

Second, consider the *quantum postulate*. This is in part the fact that Planck's constant is nonzero. It is also more than this. The quantum postulate leads to the embodiment in physics of another of Bohr's metaphysical themes – that of the inseparability of subject and object, of observer and observed, the feature of Wholeness exhibited by quantum phenomena.

Bohr says, with his usual awkwardness, that the quantum postulate

> attributes to any atomic process an essential discontinuity, or rather individuality, completely foreign to the classical theories and symbolised by Planck's quantum of action.[12]

Think of the decay of a hydrogen atom whose electron drops from an excited state to its ground state. A photon of a certain energy is given off and the electron loses an equivalent amount of energy. The whole process cannot be subdivided, you never have half a photon given off. The process has an entirely nonclassical discontinuity, the magnitude of which is dependent upon Planck's constant, as this determines the allowed energy levels of the electron. A similar wholeness appears in any interaction with a measuring apparatus as the interaction must involve an interchange of energy. The quantum postulate

> implies that any observation of atomic phenomena will involve an interaction with the agency of observation not to be neglected. Accordingly, an independent reality in the ordinary physical sense can neither be ascribed to the phenomena nor to the agencies of observation.[13]

There is a story by the Danish writer Poul Martin Moller which, it so happened, was required reading for every member of and visitor to the Institute of Theoretical Physics in Copenhagen. In the story, the hero describes what seems to him to be the paradox of thinking. In thinking we cannot control our thoughts. They just happen to us. We are a spectator. Yet we *are* the thinker. The spectator is also actor. Similarly, in quantum theory the observer and the observed become one. There is no dividing line between them. Like a blind man finding his way around a room with his stick, the stick is part of the man but also part of the room and there is no dividing line between man and room.

So the quantum postulate, the fact of the finite quantum of action, implies that the quantum system and the measuring apparatus do not, in Bohr's words, have 'an independent reality'. It is not, or presumably not, that the quantum world does not exist. But it does not have an *independent* existence in the following sense. Quantum theory deals with wholes, phenomena. The dynamical variables of position and momentum and energy, which in classical physics are properties of the microphysical system, are to be thought of in quantum mechanics as properties of the whole phenomenon. They are relational. But the non-dynamical quantities of mass, number of particles (we are thinking of nonrelativistic quantum mechanics) are properties of the quantum system. Therefore you can say how many quantum systems there are in a beam and what the mass of each system is.

The dynamical variables of quantum theory do not describe quantum systems as such. They describe the whole which consists of quantum system plus experimental measuring apparatus. Bohr says, especially in his post-EPR writings, that quantum theory describes *the phenomenon,* a whole which consists of quantum system, a measuring system, and their interaction but that this way of speaking must also be misleading since these three items cannot be separated out of the whole. This makes Bohr an *antirealist.* Our best description of the world does not lead to propositions which are true or false and whose truth or falsity is independent of the means we have for determining them.

Bohr's emphasis on the wholeness of quantum phenomena, and the consequence that descriptions of the quantum world which violate it must be false leads Bohr into an ambivalent view of the existence of separate quantum systems. Of course, there are individual electrons, in a sense. Thus in the Como lecture of 1927 Bohr tells us that 'the individuality of the elementary electrical corpuscles is forced on us by general evidence' but also that 'isolated material particles are abstractions'.[14] It is a measure of the obscurity of both the idea of wholeness and of Bohr's writing itself that it is hard to get a straight answer from him to questions like: are electrons real?

These are not very clear, and certainly not straightforwardly positivist ideas. I should also stress that it is also just one interpretation of Bohr. Perhaps we should say, as some philosophers do, that Bohr was more a Kantian than a positivist. On this interpretation we would have Bohr say that there is indeed a world of noumena behind the world of appearances, but the best we can do is to represent it by 'abstraction' which leads us astray, and into the paradoxes of the quantum theory. Of course, we are physicists and so we must represent it, but we must also always be on our guard against failing to recognize the limitations on our means of representing the world of noumena. One thing we must recognize, according to Bohr, is complementarity.

The complementarity of what?

The quantum postulate limits the scope for a classically complete description of phenomena. How does Bohr express this limitation? It is his view that the *principle of complementarity* gives the qualitative answer and the *indeterminacy relations* the quantitative answer.

Bohr tended to apply the idea of complementarity to several different categories of thing. In his writings before 1935, he associates it with the representations we make of those abstractions, individual quantum systems. Thus in the Como lecture in which he launched complementarity he says

> The very nature of the quantum theory forces us to regard the space-time coordination and the claim of causality, the union of which characterises the classical theories, as *Complementary* but exclusive features of the description, symbolising the idealisation of observation and definition respectively.[15]

The 'space–time coordination' is the description of a quantum system as having determinate positions at different times. However, a free quantum system can be localized only by forcing it to jump into states of well-defined position. If left alone it will immediately evolve into a state of ill-defined position. So either causality or well-definedness has to go.

At other times prior to 1935, Bohr speaks of the 'complementary pictures' of wave and particle, both of which are essential in our description of phenomena but which cannot be applied simultaneously to the same phenomenon.

In speaking of 'complementary pictures' Bohr touches a nerve which has disturbed many physicists who are good and, despite modern physics, sometimes unreconstructed visualizers. But Bohr is not, or need not necessarily be taken to be, speaking of pictures in any literal sense. Bohr is, or can be read as being, more concerned with what quantum systems actually are, waves or particles. The expression 'complementary pictures' should then be thought to be synonymous with the blander 'complementary descriptions'. Talk of pictures in the philosophy of physics is highly likely to mislead. This is a point on which it is worth expanding.

Theoretical physics consists in making representations of parts of the physical world and of reasoning about those representations. The representations taken together make up our 'picture' of the physical world, or at least that part of it which physics embodies. Of course, these representations need not be pictorial; in present-day thinking they tend not to be. They can be collections of sentences. These sentences include some mathematics, some sentences expressing new, unfamiliar 'theoretical' terms, and also sentences expressing interpretations of the mathematics.

We must ask: How are we to understand these representations? If they are sentences how do we connect them to the world that we experience and manipulate? In asking these questions we encounter one of the large and fundamental questions in the philosophy of science.

One answer consists in asserting that a necessary feature of explanation is reduction to the familiar, and that explaining the meaning of theoretical terms in physical theories involves connecting them with the familiar via pointing at things in the world, and via picturing. Physicists draw pictures and diagrams when they explain phenomena, and necessarily so. Pictures offer a way out of the circle of words and mathematical symbols that theoretical physics would otherwise be. Evidence for the view comes from the fact that physicists tend to be good visualizers, and that physicists distinguish the physics from the mathematics in a theory.

Therefore there is more to physics than applied mathematics, and pictures and picturable models (these may not be the only kinds of model) play an important role in making a theory intelligible. When a theory which resists visualizing comes along, it must be unintelligible. One way of expressing the difficulty of quantum mechanics is therefore to say that unlike classical mechanics it is unpicturable, or that its account of any given phenomenon demands the use of complementary pictures.

There is a good deal wrong with this line of reasoning. First, though pictures are often used in explanation they need not function in any essential way. They need not be indispensable, except in the sense that they are the only means we have for jogging ourselves into understanding. For, secondly, pictures are very limited in their expressive power. How can one express the difference between the continuity of space and its mere density in pictures? A picture needs to be supplemented by a context, a description, if it is to be meaningful. As Wittgenstein puts it

> I see a picture; it represents an old man walking up a steep path leaning on a stick. – How? Might it not have looked just the same if he had been sliding downhill in that position?[16]

Picturing, and the use of pictorial models, is an important part of the activity of doing physics. However, just as they are suggestive, so they may be misleading. There is nothing that can be drawn or visualized that cannot be said, or described mathematically. A physical theory that resists picturing may be disconcerting, but is not in itself a cause of crisis. The trouble with quantum mechanics is that we have no account of what is true and false of quantum systems. We have a theory which predicts the results of measurement but does not tell us what is true of what there is.

So much for pictures.

But there is a third, and ultimately most intriguing category to which complementarity applies, at least in Bohr's post-EPR writings. *Phenomena* can be mutually complementary. Thus, phenomena are defined by different concepts, correspond to mutually exclusive experimental arrangements, and can be unambiguously regarded as complementary in the sense that, taken together, they exhaust all well-defined knowledge about atomic objects.

The 'phenomenon', in this usage, means the whole experimental arrangement. The individual quantum system on which a measurement is performed is, in Bohr's view, an abstraction. We cannot assign properties to it as such. We can only assign properties to it in the context of a measurement. This is not to say that the measurement 'creates the property' (say, of having such and such a position). Nor is it to say that the position of the electron is a logical construction from, or is reducible to, statements about the measuring apparatus. Rather, talk of the position of an electron has sense only in the context of an experimental arrangement for making a position measurement.

The thesis of the essential indivisibility of the quantum phenomenon is not a piece of idle holistic metaphysics in Bohr's philosophy, though metaphysics it certainly is. It is a piece of working metaphysics. It plays a decisive role in Bohr's response to EPR. Just how the idea of complementary phenomena can be put to work will become clear in Chapter 5.

Where should one place the Copenhagen interpretation in that simple dichotomy of individual system versus ensemble interpretation? If Bohr is to be taken literally the answer must be nowhere. Both types of interpretation are atomistic and antiholistic. One can read Heisenberg as offering an individual system interpretation. But if individual systems really are abstractions, then Bohr's version of the Copenhagen story is neither 'individual system' nor 'ensemble', but is rather a piece of alternative metaphysics meant to make sense of an otherwise unintelligible piece of physics.

Finally, does complementarity admit of degrees? Is there a measure of it? Bohr's answer seems to have been 'yes', and that the indeterminacy relations provide the measure of it.

The indeterminacy relations

What are the indeterminacy relations? Or, more precisely, where do they come from?

There are in fact two sorts of 'derivation' of the indeterminacy relations. The first sort is a genuine derivation and employs the formalism of quantum mechanics. It defines δx and δp as *standard deviations* of x and p. The first derivation of

$$\delta x \delta p \geqslant h/4\pi$$

of this type was given by H. P. Robertson in 1929, and a version appears in Heisenberg's Chicago lectures of 1930.

A second type of informal argument employs the equations $E = h\nu$

and $p = h/\lambda$, bits of classical wave optics and various thought-experiments. The treatment Heisenberg gave to the γ-ray microscope thought-experiment in his 1927 paper was of this type, as were his subsequent treatments of a variety of thought-experiments.

The derivation of the indeterminacy relations from the formalism makes them difficult, though not impossible, to apply to the individual system. How can a standard deviation apply to an individual? We usually think of statistical concepts as applying only to ensembles, and not to the individual case. So this derivation seems at first sight to be biased towards an ensemble interpretation, not necessarily of quantum mechanics as a whole but certainly of the indeterminacy relations themselves. In a frequentist interpretation of probability, one thinks of the δx and δp as referring quite simply to dispersions of the *results of measurements,* each measurement made on a different member of an ensemble.

Informal derivations do not have this bias towards an ensemble interpretation. Bohr naturally resisted an ensemble interpretation of the indeterminacy relations, and thought that they applied to the classical *concepts* which form part of phenomena.

Bohr's view is that the indeterminacy relations apply to each and every phenomenon. Thus in any phenomenon the concepts of position and momentum will be partially defined and only to within the limits set by the inequality

$$\delta x \delta p \geqslant h.$$

This indeterminacy may well be expressed in a dispersion of the results of many measurements made with a given phenomenon. But that is not what it *means*.

As an alternative to Bohr's metaphysical view of the indeterminacy relations, one might try to apply them to an individual system. However, it is difficult to see how this could be done. The standard deviations δx and δp cannot reasonably be taken to refer to the total spreads in position and momentum for the individual system because of the complementarity theorem. The idea that they refer to total spreads has unintuitive consequences, in any case. Imagine two quantum systems, say neutrons, with total spreads of position and momentum in the x-direction of δx_1, δp_1, δx_2, and δp_2. These satisfy

$$\delta x_i \delta p_i \geqslant \frac{\hbar}{2} (i = 1,2).$$

Put them on top of one another and you would expect that the minimum spread in position would be

$$\text{maximum } (\delta x_1, \delta x_2)$$

and similarly for momentum, the minimum spread would be

$$\text{maximum } (\delta p_1, \delta p_2).$$

The product of these, in general, will be greater than the actual minimum since the two-neutron system will have greater mass than a single neutron. Therefore for a given spread δx its maximum spread δp may be *less* than either δp_1 or δp_2, and certainly less than their maximum. Put another way, one can say that the indeterminacy relations have a sort of wholeism built into them, which defies a simple individual system picture.

The final option is to view the indeterminacies as applying to the *propensities* of experimental set-ups, whatever they are. What propensities are, we leave to Chapter 5. It will turn out that this option has more in common with Bohr's Copenhagenism than might at first sight appear.

5

The Copenhagen interpretation (II): Einstein versus Bohr

The debate between Bohr and Einstein on the consistency, completeness, and finality of quantum theory ranks as one of the most significant philosophical debates of the twentieth century. It was a philosophical debate not in the sense that it had a considerable impact on the relatively technical field of the philosophy of physics. It was philosophical because at its root there lay two rival conceptions of physical reality and of our capacity to comprehend physical reality.

With the possible exception of the Leibniz–Clarke correspondence in which Leibniz and Clarke – with the latter acting as Newton's frontman – pressed their respective conceptions of space and time in the early part of the eighteenth century, the Einstein–Bohr dialogue has no parallel in the history of physics. Unlike the Leibniz–Clarke correspondence, the Bohr–Einstein debate contained no bitterness and was not based on any personal rivalry.

The debate was continuous from about 1927 and lasted right up to Einstein's death in 1955. For the contemporary philosopher of physics the most important episodes took place in the years between 1927 and 1935. However, in one respect at least the differences between Bohr and Einstein predated quantum mechanics itself.

One is tempted to view the disagreement between Bohr and Einstein as the opposition of Bohr the revolutionary and Einstein the conservative, and in his best and clearest account of it Bohr[1] certainly writes as if that were so. One is liable to see Bohr as reading into quantum theory new and profound epistemological and metaphysical consequences, and to see Einstein as the reactionary who had a hankering for an outdated classical ontology. The truth is more complex. It is better to think of Bohr and Einstein as both radical and conservative, but in different ways. Thus Einstein was the radical who would sacrifice classical electromagnetic theory while hoping to recover classical causality in some way. Einstein's conservatism was expressed in his desire for determinateness and determinism. Bohr, on the other hand, was the radical who wanted to sacrifice causality while conserving as much as he could of classical electromagnetism.

The differences between Einstein and Bohr begin with Bohr's and Einstein's attitude to the latter's hypothesis of free light quanta. Einstein's explanation of the photoelectric effect preserves causality at the cost of limiting the general validity of electromagnetic theory. In the years before the discovery of the Compton effect in 1923, Bohr sided with the conventional opinion of the day, which regarded the free light quantum as a fiction. As late as 1921 Bohr, in his address to the third Solvay Congress, was using the stock objection to particle interpretations of quantum mechanics as an argument against the free light quantum, namely that

> [s]uch a concept seems, on the one hand to offer the only possibility of accounting for the photo-electric effect, if we stick to the unrestricted applicability of the ideas of energy and momentum conservation. On the other hand, however, it presents apparently insurmountable difficulties from the point of view of the phenomenon of optical interference . . .[2]

Why was Einstein so attached to the free light quantum, or as we now say, the photon?

Recall that Planck construed the quantum of action to govern only the *exchange* of energy between matter and radiation and that freely propagating radiation should be governed by the equations of the classical electromagnetic field. The majority of physicists took the same view of the photoelectric effect. Indeed it was only the discovery of the Compton effect in 1923 in which freely propagating radiation exhibited particle behaviour that convinced most physicists of the need for light quanta in transit.

Much of Einstein's effort in the old quantum theory had been devoted to defending his free light quantum hypothesis. In 1910 he used Planck's law to show that the expression for the energy fluctuations about the average for frequencies between ν and $\nu + \mathrm{d}\nu$ for a black body were the sum of two terms, one a *wave* term and the other a *particle* term which Einstein interpreted as confirming the existence of free light quanta. Later, in 1916 in a classic paper, he derived Planck's law from simple assumptions about the processes of absorption and emission of free light quanta.

Bohr continued to adhere to an anti-free light quantum picture of light to the last possible moment. In his watershed paper of 1924 with Kramers and Slater[3] (hereafter BKS), a paper which represents the last piece of original semiclassical atomic physics before the revolutionary development of quantum mechanics, Bohr developed the idea of dropping the 'unrestricted applicability' of the laws of conservation of energy and momentum into the basis for a new program in atomic physics. BKS imagine atoms to be excited not by their direct interaction with individual photons but rather with a continuous background field. Though the field changes continuously as classical electromagnetic theory demands, the atoms which interact with it change discontinuously. The physical properties of the field are rather obscurely imagined to be determined locally by the many atoms nearby. The individual interactions of the atoms with the field are not to be governed by strict conservation laws. These will hold *only statistically* for a large number of interactions and not for each individual interaction.

BKS is a remarkable paper, a landmark in the development of quantum mechanics, and a fine example of Bohrian 'physics without mathematics', containing as it does only one equation, and that, $E=h\nu$. It is a landmark because it marks the end of attempts to modify classical physics by simply overlaying the classical ontology with quantization plus indeterminism. The BKS idea was quickly refuted by the experiments of the other quantum heroes of experiment Bothe and Geiger who showed that the conservation laws were obeyed *in each individual interaction,* a fact which Bohr immediately admitted, as for example in his Como lecture. Bothe and Geiger had developed coincidence techniques which were able to show that in the Compton effect the secondary photon and the knocked-on electron are produced within 10^{-3} sec. of one another, which they would not be if photon and electron were not directly connected. (Incidentally, more recent experiments have reduced the time interval to 10^{-11} sec.) Similar experiments using a cloud chamber were performed by Compton and Simon at about the same time and with the same result.

The failure of BKS was an important stimulus to quantum mechanics proper. The experiments of Bothe and Geiger comprise a difficulty for any interpretation of quantum mechanics as a kind of generalized and indeterministic classical mechanics of which Popper's interpretation (which we examine later in this chapter) may be taken as an example. Einstein was critical of BKS when it was published, his principal criticism being that to abandon strict causality was to free oneself of an essential constraint on theorizing in physics. For Bohr, the essential constraint lay with the other classical idea of continuity which the free light quantum hypothesis ignored.

Being on opposite sides over the photon was a mere preliminary to their larger debate which followed the Solvay Congress of October 1927.

What we now call the Copenhagen interpretation became a subject of public discussion in the weeks following Bohr's Como lecture. The fifth Solvay Congress was convened under the title *Electrons and Photons* but the matter of real interest was the correct interpretation to be put upon the new quantum mechanics.

In one session Bohr repeated the contents of his Como lecture and Einstein replied during questions in which he distinguished two *Viewpoints*. According to Viewpoint One, a wave-function or state-vector should not be regarded as describing an individual system but only an ensemble of systems. According to Viewpoint Two the state-vector should be taken to be the state-vector of an individual system and as embodying a complete description of the system.

Einstein seems to have thought that the results of Bothe and Geiger support Viewpoint Two. But this was to confuse the strict application of the conservation laws to each individual interaction with the applicability of quantum mechanics to each system involved in the interaction. It would be quite consistent, at least prima facie, to hold strict conservation in each individual system interaction with an ensemble interpretation of quantum mechanics, as a proponent of a deterministic hidden-variables theory might.

But what was especially unacceptable in Viewpoint Two, in Einstein's opinion, was that it contradicted the special theory of relativity in a particularly unacceptable way. When the position of a free particle is measured its state-vector collapses discontinuously, *faster than light.* Think of the one-slit experiment performed with electrons. Just before an electron is localized on the screen it will spread right across it. When it impinges on the screen the state-vector collapses to be contained in a very small region on the screen.

This way of disposing of Viewpoint Two seems to assume that according to it, one must identify a system with its state-vector. However, there are already good reasons for not doing so, reasons which we enumerated against Schrödinger's interpretation of wave mechanics. This objection of Einstein's to the Copenhagen interpretation was thus inconclusive. An individual system interpretation need not identify the spatial extent of an individual system with the extent of its wave-function.

One can think of the Bohr–Einstein debate in the period 1927–35 as being made up of two phases. In the first, lasting at least until the sixth and next Solvay Congress of 1930, Einstein tried by means of thought-experiments to show that the Bohr–Heisenberg interpretation of the indeterminacy relations was inconsistent with another part of the Copenhagen interpretation, namely the thesis of completeness. Einstein argued that with sufficient ingenuity one could construct thought-experiments in which the indeterminacy relations could be circumvented and that one could arrive at a more complete description of an individual quantum system than the indeterminacy relations seemed to allow.

When this strategy failed, or seemed to have failed, Einstein adopted the more direct tactic of trying to show that quantum mechanics is incomplete, that thought-experiments could be constructed which show that quantum systems must be carrying information which, though not described by quantum mechanics, nevertheless determines how they will behave under measurement. This second phase of Einstein's attack on the Copenhagen interpretation culminates in the EPR paper and in Bohr's reply to it.

Einstein on the indeterminacy relations

The orthodox view, the view of Heisenberg for example, and the apparent view of Bohr in 1927, was Einstein's Viewpoint Two, that quantum mechanics gives a complete description of the individual quantum system, that quantum mechanics says everything there is to say about it. As a corollary, the indeterminacy relations expressed quantitatively the limitation on any description of a quantum system employing classical concepts.

If Einstein could show that the indeterminacy relations could be violated when one took them to apply to the individual system then he could refute the Copenhagen interpretation. Of course, to violate the indeterminacy relations is not to refute quantum mechanics itself but only the metaphysical constructions that Bohr and Heisenberg had put on it. So Einstein set about the indeterminacy relations with thought-experiments. His first examples involve diffraction using one and two slits. The most subtle of these was a variant of the two-slit experiment, and he put it to Bohr at the first Solvay Congress in 1927.

Imagine a beam of electrons emerging from a source and moving in the x-direction towards the usual diaphragm in which two slits have been cut. The line normal to the slits and lying in the plane of the diaphragm is the y-direction. This time the diaphragm is suspended by a spring in such a way that the *impulse* (the total change in momentum) delivered to the diaphragm due to its interaction with each electron may be measured. Then, in the usual way, a screen is set up behind the diaphragm and the position of each electron hit on the screen may be marked.

If you know the impulse any given electron delivers to the diaphragm then you can calculate its momentum in the y-direction after it negotiates the diaphragm and this will presumably be its y-momentum when it hits the screen. But if you know its point of impact on the screen you can calculate which of the two slits it passed through, given that you know its y-momentum when it hits the screen. If you can identify which of the two slits it passed through, and these can be made as narrow as you like, you can measure its y-position and its y-momentum as accurately as you like.

In this account Einstein assumes the diaphragm to be unproblematically *determinate* and he does not apply the indeterminacy relation to the diaphragm. In his reply to Einstein, Bohr does exactly that. Extending the application of the indeterminacy relations was a typical ploy of Bohr's and it was one in which he showed himself the supreme master of intuitive physics.

Let the angle subtended at the screen by the distance between the slits be ω. Then the uncertainty δp in the momentum of the electron as it strikes the screen roughly opposite the slits will be of the order of ωp where p is the actual momentum of the electron. So that

$$\delta p = \omega p = h\omega/\lambda$$

where λ is the de Broglie wavelength of the electron, assumed to be represented by a state-vector which is a plane wave like the example g of Chapter 2.

However, the measurement of an electron's position on the screen will involve an uncertainty δx where

$$\delta x = \lambda/\omega,$$

this according to classical wave optics. Therefore the product $\delta x \delta p$ will be of the order of h. So you cannot measure the impulse given to the diaphragm with sufficient accuracy to determine which slit the electron went through.

Note that Bohr is not *deriving* the indeterminacy relations. He is simply *defending* them against a particular sort of attack, namely that they may be violated given certain assumptions: that the electron really does have simultaneously determinate position and momentum. His defense is that Einstein has not applied the indeterminacy relations sufficiently thoroughly. In defending the indeterminacy relations against Einstein Bohr is free, from a purely logical point of view, to use Einstein's assumptions even though he may think them false.

It may seem odd that Bohr should apply the indeterminacy relations to a classical object, the diaphragm. But this does illustrate one aspect of Bohr's version of the Copenhagen interpretation and it is one which marks him off from many orthodox expositors of that interpretation. There is, in Bohr's view, no 'cut' between the quantum world and the classical world. The classical world is a quantum world. Quantum mechanics must employ our genetically prior classical concepts which are not fully applicable in the quantum world. But they are not *fully* applicable in the classical world either, which is only classical because the extent of the inapplicability of classical concepts, as revealed by the indeterminacy relations, is undetectable at the classical level.

Rebuffed on this occasion, it was clear to Einstein that he needed to find a more conclusive counterexample. Bohr held that quantum mechanics and special relativity had in common a philosophical limitation of classical concepts – simultaneous position and momentum in the case of quantum mechanics, frame-independent simultaneity in the case of relativity. Heisenberg drew the analogy explicitly in his γ-ray microscope paper, an analogy which Bohr expected to appeal to Einstein. Of course, it wasn't so. Therefore, and ironically, Einstein set out to use relativity against Bohr.

Einstein's problem was this. How can you measure the energy of a photon *and* the time at which it is released from some apparatus with a precision sufficient to violate the indeterminacy relation

$$\delta E \delta t \geqslant h?$$

Einstein presented his *photon-in-a-box* thought-experiment to Bohr at the sixth Solvay Congress of 1930 which attempts this in the following way. (By the way, the energy–time uncertainty relation is problematic in itself. Time is not a quantum-mechanical observable. When one knows the 'time' of a system one knows nothing about it. The usual derivation of the uncertainty relation is therefore going to fail. One interpretation has it that a measurement of energy which takes only δt must be uncertain by an amount δE but its interpretation is still a matter of controversy.)

Nevertheless, here is Einstein's thought-experiment. Imagine a box full of photons whose energy, if they escape from the box, is known precisely. Imagine that the box has perfectly reflecting walls and a small hole closed by a shutter. The shutter can be made to open for a time short enough to emit a single photon. The whole apparatus of box and shutter is placed on a weighing scale. If a photon escapes from the box, the mass of the apparatus will decrease, the photon carrying off an energy E associated with a mass mc^2, where m is the decrease in the mass of the apparatus.

The aim of the experiment is to determine the color – that is, the frequency and hence energy – of the escaped photon by weighing the box. The energy E equals mc^2 which equals $h\nu$, where ν is the frequency of the photon. Open the shutter and note with as great an accuracy as you like the small time δt for which it was open. Weigh the box to see if you have lost a photon. You have lost a photon if the box has decreased in mass. Imagine that you have arranged the intensity of light in the box and the size of the hole so that you would have to be very unlucky to release more than one photon from the box at a time. If there is a decrease in the mass of the box you can measure it as accurately as you like. So, it seems, you can measure δt, given by the shutter speed, and δE as accurately as you like.

But Bohr came up with a strange and ingenious reply which appealed to the fact that clocks slow down in a gravitational field. To be exact, Bohr employed the Einstein's *general relativistic red-shift formula* which gives the relative slowing down of a clock moved through a gravitational potential difference $\delta\phi$

$$\delta t/t=\delta\phi/c^2.$$

He argued as follows. First, and as usual, apply the indeterminacy relations to the apparatus *thoroughly*. Determining the mass of the apparatus, Bohr argued, is really a matter of determining a weight. The weighing procedure must involve something equivalent to a spring balance or some scales or something which must move through a gravitational field. To determine the zero position to an accuracy δx requires a minimum uncertainty in the momentum of the apparatus of δp and these, Einstein seems now to admit, satisfy

$$\delta p\delta x\geqslant h.$$

Let the time it takes to do the weighing be t. If the process is a genuine weighing of the box and a determination of the mass of the photon in a time interval t, then δp must be less than the total impulse – the total change in momentum – given by the loss of the photon to the scales in the time t. So that

$$\delta p<tg\delta m \tag{1}$$

where g is the acceleration of gravity and δm is the decrease in mass of the apparatus due to the loss of the photon.

General relativity tells us that our measurement of t is subject to the following uncertainty δt

$$\delta t=tg\delta x/c^2 \tag{2}$$

since the clock must move through the gravitational field a distance whose uncertainty is δx. Therefore, eliminating tg from (1) and (2)

$$\delta t>\delta p\delta x/(\delta mc^2)$$

and with $\delta E = (\delta m)c^2$ from relativity, and $\delta p \delta x \geqslant h$ from the indeterminacy relations, one has

$$\delta E \delta t > h.$$

Bohr's argument is again that a *consistent and thorough* application of the indeterminacy relations yields no contradiction. He argues that Einstein has ignored the uncertainties of the weighing process which induce an uncertainty in the energy of the photon inversely proportional to the time the weighing takes.

Einstein's thought-experiment is open to at least two implementations.

One can leave the shutter open and wait for a photon to escape. That one has escaped is indicated by the loss of mass of the box. In this case there are the limitations that Bohr describes in the energy of the photon and in the time at which the photon was emitted.

On the other hand, one can open the shutter at a given time for as short a time as one likes, and then one can weigh the box to see if a photon has been emitted. One can take as long as one likes to weigh the box, thus reducing the uncertainty in the energy of the emitted photon. So, if one is lucky and a photon does escape, it *seems* that one can know the energy of the photon and the time at which it escaped with as much accuracy as one likes. But this neglects, among other things, the impact of the very fast moving shutter on the escaping photon. Presumably this interaction with the shutter would be uncertain and would give the box a slight oscillatory motion. Measuring the size of the motion would again involve an uncertainty, and so we are back to the indeterminacy relations. A complete analysis would be difficult to give, illustrating the fact that when you start applying the indeterminacy relations it is not always clear where you should stop. Certainly, Bohr was not happy with the way he had seen off Einstein, and the 1930 Solvay Congress began only a brief hiatus in the exchanges between Bohr and Einstein.

It might also seem odd that Bohr should appeal to relativity in order to defend what is surely a logically independent theory. If the Copenhagen interpretation is defensible, so you might argue, then surely it should stand up as an interpretation of quantum mechanics on its own, without appeal to any other physics theories. But notice that Einstein's thought-experiment is itself constantly appealing to relativistic ideas through its use of the photon. Einstein's thought-experiment is essentially relativistic. Note also that the red-shift formula can be derived as an approximation from special relativity alone.

The photon-in-a-box has generated a good deal of heat. It is certainly one of the most obscure thought-experiments that Bohr and Einstein discussed. The indeterminacy relations $\delta E \delta p \geqslant \frac{1}{2}h$ are troublesome in themselves. Unlike p, x and E, t is not an operator in quantum mechanics. The interpretation of $\delta E \delta t \geqslant \frac{1}{2}h$ is even less clear than that for $\delta p \delta x \geqslant \frac{1}{2}h$. Bohr was himself not happy with his victory over Einstein. But it was enough to convince Einstein that he should change tack and aim directly at the completeness of quantum mechanics. This was the target of his most powerful criticism of quantum mechanics – the puzzle of the EPR thought-experiment.

The Einstein–Podolski–Rosen argument

Recall that Einstein accepted the ensemble interpretation of quantum mechanics, Viewpoint One. If quantum mechanics could describe ensembles but not individual systems then it must be incomplete. To persuade someone who held Viewpoint Two Einstein needed an argument that assumed an individual system interpretation, and this EPR was intended to provide.

Imagine two quantum systems which either *interact* or *are created* in such a way that their properties are correlated by conservation laws. Call the two quantum systems I and II. Now suppose that because of the correlation you can infer the value of an observable **A** for II when you have made a measured **A** on I. Suppose similarly that you can infer the value of **B** for II when you have measured **B** on I. Let **A** and **B** be totally incompatible, that is for all ψ in the domain of **A** and **B**

$$[\mathbf{A}, \mathbf{B}]\psi \neq 0.$$

You can predict either the value that **A** has or the value that **B** has for II, depending on your choice of measurement on I. The systems I and II can be separated in space. They may interact or be created together and then drift apart and the measurement on I can be made when they are far apart.

Assuming the *locality principle* that the choice of measurement on I cannot affect the properties of II, it seems that system II must carry away information which determines how it will respond to measurements of both **A** and **B.** But if **A** and **B** are totally incompatible there will be no state-vector which for system II conveys this information both about **A** and about **B.** The information must be there to be described. Therefore quantum mechanics is incomplete.

Some practical details. Among the possible correlations for *electrons* I and II are

(*a*) that their total momentum and their center of mass are fixed
(*b*) that their total spin is zero.

Case (*a*) corresponds to that of two identical classical particles rebounding in opposite directions and with the same speed from a collision. It is essentially the case EPR consider, though they do not say what caused the correlation of the properties of the systems. Their **A** and **B** are position and momentum. Furthermore, they give us a wave-function for I + II in one dimension, the *x*-direction, which is an eigenfunction of the sum of the momenta $p_I + p_{II}$ and of the difference of the positions $x_I - x_{II}$. Their wave-function is such that measuring *p* for I enables us to predict p_{II}, and similarly such that measuring *x* for I enables us to infer x_{II}. But there is no wave-function either for I + II as a joint system, or for II alone, which has simultaneous definite values for p_{II} and x_{II}. Therefore *either* there is action at a distance, on II by the measurement on I, *or* quantum mechanics is incomplete.

EPR give us two criteria, the first a *sufficient condition* for the existence of an 'element of reality', the second a *necessary condition* for the completeness of a physical theory. The first is a sufficient condition for 'ontological commitment to a physical quantity':

> [i]f, without in any way disturbing system, we can predict with certainty (i.e. with a probability equal to unity) the value of a physical quantity, then there exists an element of reality corresponding to that quantity.[4]

This criterion cannot be a necessary condition for the 'reality of a quantity' since there may be, and presumably are, quantities which we do not even know about, and whose values we cannot predict with certainty. Applying this criterion is not without its problems. How do we know we are truly predicting the value of a quantity? We might have a false theory complete with criteria for determining values (with probability unity) of physical quantities which simply do not 'exist' in nature at all. We trivialize the criterion if the theory has to be true. What we must say is that *we* should be committed to elements of reality corresponding to dynamical variables we *think* we can predict with certainty. If we can predict with certainty there must be something in the world making our predictions come true, barring miracles.

EPR's second criterion deals with the completeness of a physical theory. It is a necessary condition, they say, for the completeness of a physical theory that

> [e]very element of the physical reality must have a counterpart in the physical theory.[5]

It would be more than optimistic to think that any of our physical theories is complete in EPR's sense, or that any theory ever will be, or perhaps that any intelligible theory ever could be. To think that we could possess a complete physical theory is to think that the world is completely describable from top to bottom and that we could know what the true ultimate description of the world is. If knowledge of completeness is beyond us, knowledge of incompleteness need not be. This is what the EPR thought-experiment provides: EPR try to show that *relative to the locality assumption,* quantum mechanics must be incomplete: that is, there must be features of the quantum world which quantum mechanics fails to specify.

As a matter of fact, the particular thought-experiment that EPR invented has several disadvantages. In the contemporary discussion of nonlocality it has been superseded by a simpler example due to David Bohm.[6] The EPR example would be difficult to realize in practice, and it employs the dubious machinery of δ-functions. Bohm's example (call it EPR-B) which follows is that of case (*b*) above and, with photons instead of electrons and in a slightly more complex form, it forms the basis of the experimental tests of nonlocality performed during the 1970s.

Imagine the decomposition of (say) a diatomic molecule, Bohm's example, or a pair of correlated electrons, in either case the total spin of the systems being zero – the *singlet state* – into two atoms (or electrons) each of which has spin $\frac{1}{2}h$. We represent the singlet state of the molecule (electron pair) by the state-vector

$$|\mathrm{I, II}\rangle$$

and the spin states of the separate atoms by

$$|+(-),d,\mathrm{I}\rangle \text{ and } |+(-),d,\mathrm{II}\rangle$$

where the spin 'up' ('down') states are represented by the $+(-)$ and d is any given direction, so that

$$|\mathrm{I, II}\rangle = (1/\sqrt{2})\{|+,d,\mathrm{I}\rangle|-,d,\mathrm{II}\rangle - |-,d,\mathrm{I}\rangle|+,d,\mathrm{II}\rangle\} \quad [***]$$

Notice the '$-$' on the right-hand side of the equation for $|\mathrm{I, II}\rangle$. This antisymmetrizes $|\mathrm{I, II}\rangle$ making it change sign under exchange of I and II, as required by the exclusion principle for spin $\frac{1}{2}$ particles.

The atoms (electrons) I and II separate after the decomposition. A '$+$' ('$-$') result for a measurement of spin in a direction d on I tells us that II would now be found to have spin in the opposite d direction '$-$' ('$+$'). A difference choice of d would yield exactly the same result. But no state-vector gives determinate values of spin in two different directions. As for the EPR, so for EPR-B. It must be that system II is predetermined to respond to spin measurements in every direction in a way not captured by the quantum-mechanical description. Unless, that is, there is action at a distance.

Here we see a recurring theme of Einstein's objections to the completeness of quantum mechanics: that completeness would violate locality, a fundamental assumption of the theories of relativity. It will turn out, as we shall see, that quantum mechanics is a nonlocal theory even if it is, as EPR thought, incomplete but statistically correct. Ironically, this was discovered thirty years after the EPR paper by a deeper analysis of the EPR thought-experiment.

The source of the apparent connection between particles I and II in the EPR thought-experiment is the fact that their state is described by a state-vector [***] whether they are close together or millions of miles apart. Surely it would be reasonable to think that as they travelled they decayed into an uncorrelated pair whose joint state should be represented not by a state-vector but by a 'mixture' such that the state of II would not be affected by measurements on I. This suggestion has two difficulties. First it would imply a breakdown of some conservation law for the individual interaction, the violated law being conservation of angular momentum in the EPR-B example. There is no evidence of such a breakdown for individual interactions and much evidence against it. Second, a classic experiment by Wu and Shaknov[7] using photons whose polarizations are correlated by their being produced from collisions of positron–electron pairs confirms that the polarizations of the two photons are indeed opposite as implied by the state-vector description.

Thus Bohr's reply to Einstein required a subtle analysis of Einstein's presuppositions and not an objection to his physics. The result was the full articulation of Bohr's relational conception of phenomena.

Bohr's reply to EPR

Bohr's immediate reply was published, like EPR, in *Physical Review* in 1935. His argument was that there is an *ambiguity* in the EPR criterion of ontological commitment.

EPR claim that if *without in any way disturbing the system* we can predict the value of one of its dynamical variables with certainty then there must be something in the world corresponding to it. But, Bohr says, EPR do 'disturb' the overall system which incorporates I and II when they make the change from measuring **A** on I to measuring the incompatible **B** on I. EPR take the system they are describing to be the individual quantum system. But, according to Bohr, the system is really the composite entity the *phenomenon,* and this entity incorporates both the abstraction of the individual systems I and II and the experimental arrangement which performs the measurement on I. In measuring position or momentum, or spin in two different directions, you are choosing between two different phenomena. There is no question of a mechanical disturbance of the system under investigation during the last critical stage of the measuring procedure. But even at this stage there is essentially the question of an influence on the very conditions which define the possible types of predictions regarding the future behavior of the system . . . these conditions constitute an inherent element of the description of any *phenomenon* to which the term 'physical reality' can be properly attached.[8]

Bohr here provides an alternative to the dichotomy of individual-system versus ensemble interpretations. The 'phenomenon view' is perhaps closer to the individual-system view. But Bohr always writes ambiguously on the question of whether there really are individual systems to be described. There are individual systems but it also seems that they are 'abstractions'. In making a measurement of spin in one direction we employ one phenomenon. Then, in changing to another direction, we 'disturb' the original phenomenon.

The outcome of the Bohr–Einstein debate was a new metaphysics of 'phenomena' on the part of Bohr and a new strategy for criticizing quantum mechanics on Einstein's part. Neither Bohr nor Einstein conceded victory to the other, though the great majority of physicists who wanted a quiet (that is, an unphilosophical) life took the Copenhagen side, whatever they took that to be. But whether or not Bohr's metaphysics is more satisfactory than it is plausible, and whether or not Einstein's demand for nonlocality is satisfiable, will become more apparent in Chapter 8.

Ensemble interpretations

Bohrian Copenhagenism might fudge the question of the reality of individual quantum systems but most of us, and most physicists, accept them as a given item of the furniture of the world. Individual quantum systems can be counted. There are two electrons in a helium atom and there are of the order of 10^{23} helium atoms in one gram of helium gas. We know the mass and charge of an electron, of a proton, a neutron, an alpha-particle.

What general argument is there for Einstein's Viewpoint One, that quantum mechanics fails to describe the individual system and describes only the ensemble, as any ensemble interpretation would have it? The most philosophically naive argument is this: that in science the word 'probability' is always to be interpreted objectively, and this *means* as relative frequency in an ensemble.

This is not to say that the word 'probability' must always be interpreted objectively. It may be that in everyday life there is room for an interpretation which reads probability subjectively, as a measure of one's belief in a proposition, for example. But in science subjectivist interpretations of probability such as this must be ruled out as irrelevant, or so the argument goes. Therefore, since quantum mechanics deals essentially in probabilities, it must describe ensembles and not the elements of an ensemble. Exactly this view is expressed in an entertaining book by the physicist F. J. Belinfante.

When in quantum theory we assign exact values to probabilities, we consider the case of an 'infinitely' large collection of cases among which a relative frequency is calculated. The theory, therefore, deals with *ensembles*. State-vectors determine the probability distributions in ensembles: *State-vectors are properties of ensembles,* and quantum theory is a theory about properties of ensembles.[9]

Belinfante's is a simple view, much too simple in fact. It so happens that there are objective interpretations which do assign probabilities to individual events or to individual experimental set-ups. For example, Popper's propensity interpretation (of which more later) does the latter.

Therefore one must distinguish Belinfante's view from the rather weak argument in its favour. Belinfante's view – the minimal ensemble interpretation, *minimalism* for short – is attractive, disposing as it does of the paradoxes of quantum mechanics. The price it pays is a severe restriction of the scope of the theory. Take EPR as an example. EPR is intended to show that given the locality assumption quantum mechanics is incomplete when interpreted as a description of the individual system. If you hold that quantum mechanics is complete you are forced to accept nonlocality. But EPR does not demonstrate a violation of *statistical* locality. In other words, the relative frequencies of measurement results on system II cannot be altered by measurements made on I.

Similarly, minimalism disposes of the macabre paradox of Schrödinger's Cat which Schrödinger describes as follows.

A cat is penned up in a steel chamber, along with the following diabolical device (which must be secured against direct interference by the cat): in a Geiger counter there is a tiny bit of radioactive substance, *so* small, that *perhaps* in the course of one hour one of the atoms decays, but also, with equal probability, perhaps none: if it happens, the counter tube discharges and through a relay releases a hammer which shatters a small flask of hydrocyanic acid. If one has left this entire system to itself for an hour, one would say that the cat still lives *if* meanwhile no atom has decayed. The first atomic decay would have poisoned it. The ψ-function of the entire system would express this by having in it the living and the dead cat (pardon the expression) mixed or smeared out in equal parts.[10]

It would seem that after an hour the cat must be in a superposition of 'dead' and 'alive' states and must be neither dead nor alive if quantum mechanics describes it. But we know that the cat will be either dead or alive and nothing in between.

One solution to this apparent contradiction has it that the cat does evolve into a superposition of dead and alive until the consciousness of the observer who looks at the cat flips it into a determinate 'dead' or 'alive' state. Another solution – that of minimalism – is to admit the superposition but to claim that it describes only an ensemble of similarly prepared cat plus poison apparatuses and to deny that it describes the individual cat. The fact is that the superposition makes the right *statistical* predictions – proportions of dead and alive cats – for the ensemble.

The minimal ensemble interpretation sidesteps the paradoxes of EPR, of measurement, of wave–particle duality, by limiting the scope of quantum theory and by, in a sense at least, accepting incompleteness. A measurement of electron spin (for example) will then simply divide an ensemble of electrons into two subensembles, one of spin 'up' electrons and the other of spin 'down', rather than involve a collapse of anything.

About realism – the question of whether or not an individual quantum system has simultaneous exact values for all its dynamical variables – the strict minimalist should be agnostic. Therefore one can think of minimalism as an attempt to restrict or to resist the essential question of the philosophy of quantum mechanics: what does quantum mechanics tell us about the world?

However, there is a natural tendency not to rest content with full minimality. Most philosophers of physics who favour an ensemble interpretation want to append to it the idea that though, or perhaps because, quantum mechanics does not describe the individual quantum system it can have a trajectory, be essentially a classical particle. One sees this tendency in the richer ideas of the physicist L. E. Ballentine and the philosopher Sir Karl Popper.

We first discuss Ballentine, whose *Physical Review* article of 1970[11] is the classic statement of this realist extension to the minimal ensemble interpretation. These are Ballentine's main theses.

First, that state-vectors are not complete descriptions of the individual system, they are to be associated with ensembles and are the *mathematical representatives of state preparation procedures.*

Second, that a quantum ensemble is an infinite conceptual set of individual systems.

Third, that each individual quantum system has simultaneous determinate values for all its dynamical variables.

Fourth, that EPR is a paradox only for individual system interpretations.

Fifth, that the uncertainty principle is a statistical dispersion principle which applies to state preparation procedures and not to measurement and which

> restricts the degree of statistical homogeneity which it is possible to achieve in an ensemble of similarly prepared systems, and thus . . . limits the precision which future predictions for any system can be made.[12]

Sixth, that a measurement is a *selection* of a subensemble from an ensemble.

Finally, that the realist ensemble interpretation is completely open with respect to hidden variables.

If the classical Newtonian ontology is Democritean, the metaphysics of Ballentine's (and Popper's) realist ensemble interpretation is an Epicurean atomism. Like those of Epicurus whose 'atoms in the void' had an occasional, extremely small, spontaneous deflexion from the original direction of movement, Popper's atoms behave indeterministically. As in the BKS theory, the conservation laws do not apply to them individually.

The key idea with which Ballentine extends minimalism is his third thesis, that of quantum-mechanical realism. (Although Popper would say, and most would disagree, that classical mechanics itself is indeterministic.) But Ballentine (and Popper, as we shall see) must face the familiar question: how can a realist version of quantum mechanics handle diffraction and interference phenomena of which we take the two-slit experiment as archetypal?

Ballentine's answer (and Popper's) consists in giving a mechanical explanation which is due essentially to the physicist Duane.[13] The details of this mechanical account are not important. However Duane succeeded in giving a particle explanation of diffraction by a periodic crystal structure by proposing a third quantization rule in addition to

$$E = h\nu$$

and

$$p = h/\lambda.$$

Duane proposed that where l is the length of the period in the crystal lattice, the interaction of particle and crystal could result in changes in the particle momentum of multiples of h/l.

Both Ballentine and Popper[14] propose to regard the two-slit diaphragm as a periodic structure like a crystal and to provide a sketch of a mechanical explanation of the two-slit experiment along the lines of diffraction by a crystal lattice. Few writers on the philosophy of quantum mechanics seem to take this sufficiently seriously to consider the obvious objection to it which is this. There is nothing essentially mechanical about quantum-mechanical diffraction phenomena which can be performed with *lasers* or with electron microscopes. For only one of many examples see Rosa.[15]

If Ballentine's interpretation fails to handle properly the wave aspects of wave–particle, so does Popper's. But Popper has some novel technical ideas in the interpretation of probability which make his account of quantum mechanics the more interesting. In effect, Popper rejects Belinfante's appeal to the relative frequency interpretation of probability, and offers an alternative objectivist interpretation of his own.

Popper and The Great Quantum Muddle

Popper's interpretation of quantum mechanics is, like Ballentine's, a realist extension of the ensemble interpretation. But unlike Ballentine's, it exhibits, and to some extent applies to quantum mechanics, a new philosophical invention, the propensity interpretation of probability. In the propensity interpretation, you say that probability statements refer to

> some measure of a property (a physical property, comparable to symmetry or asymmetry) of the *whole experimental arrangement*; a measure, more precisely, of a *virtual frequency*.[16]

whereas you

> look upon the the corresponding *statistical statements* as statements about the corresponding *actual frequency*.[17]

Statistical statements express the results of observations, whereas probability statements describe *propensities* which are inherent in experimental arrangements. Of course, you expect the statistics to converge on the probabilities when you repeat an experiment a large number of times. Probabilities describe the propensities of the experimental arrangement. Popper is here thinking of the single experimental set-up, the experiment being performed perhaps only once. He is not thinking of propensity as a property of a collection or type of experimental arrangements. It is the bit of apparatus before you that has the propensity.

Popper's views are expressed in most detail, in fact as a list of thirteen theses, in an important paper of his entitled 'Quantum mechanics without "The Observer" '.[18] Quantum theory, he urges, is an essentially statistical theory because it was designed to solve statistical problems: the problem of black-body radiation, the problem of the photoelectric effect (more dubiously), and the Bohr theory of the atom (even more dubiously). However, statistical problems demand a statistical answer. Therefore when probability arises in quantum mechanics, it arises because quantum-mechanical problems are statistical and *not* because of ignorance. Thus the interpretation of quantum probability as a necessary limitation on our knowledge of nature is quite wrong-headed.

That which leads to unnecessary difficulties in the interpretation of quantum mechanics, namely The Great Quantum Muddle, is, in Popper's view, the attempt (of Bohr and Heisenberg in Popper's account) to assign probabilities to the individual quantum system, and this leads to the untenable ideas of wave–particle duality and complementarity. Popper's target is the more general subjectivism which he sees as corrupting modern science, or, if not modern science, the pronouncements of its leading ideologues. Quantum mechanics is inherently probabilistic, and in Popper's view this leads Bohr and Heisenberg, who are (in Popper's view) subjectivists in probability, to introduce the 'Observer' into quantum theory. Of course, the 'Observer' in Bohr's accounts need not be human. And the 'Observer' gets into quantum theory via the puzzle about quantum-mechanical measurement, specifically, the puzzle as to how it is that macroscopic reality is determinate.

Furthermore, there is no duality of wave and particle in quantum mechanics, according to Popper, except for the fact that the wave describes probabilities and the particle describes the individual quantum system, so individual quantum systems may be essentially underdetermined (Popper rejects determinism even in classical physics) classical particles. Popper insists that they are.

Popper's argument for realism derives from his critique of Heisenberg's view of the uncertainty principle. In Popper's opinion the Heisenberg inequalities do not refer to individual systems. Like Ballentine, Popper holds that they are statistical scatter relations which refer to ensembles.

> [T]he scatter relations tell us that *we cannot prepare experiments* such that we can avoid, upon repetition of the experiment, (1) scattering of the energy if we *arrange* for a narrow time limit, and (2) scattering of the momentum if we *arrange* for a narrowly limited precision. But this means only that there are limits to the *statistical homogeneity* of our experimental results.[19]

In fact, in order to be able to test the Heisenberg inequalities as applied to ensembles, we must be able to measure the values of conjugate variables for an individual system more accurately than Heisenberg allows. In Popper's view this can be done.

Given his realism, exact values of the dynamical variables are there to be measured. Consider a specific case. Measure the two positions of a particle at two times. Since you know the positions and the time difference you can work out the velocity of the particle and therefore its momentum all with an accuracy as great as you like. Heisenberg considered this possibility and rejected it, rather lamely, on the grounds that calculated momentum was of no use in making further predictions. It would be 'disturbed' by the second position measurement, so that accepting the calculated momentum value was a matter of 'taste'. The taste of Popper the realist is to take it seriously.

So what is the propensity interpretation of probability good for?

First, it makes quantum mechanics objective. Quantum mechanics describes the propensities of experimental arrangements and so there is no need to bring in 'The Observer'.

Second, it enables us to explain the paradoxes.

For example, take the two-slit experiment which Popper approaches via a pin board down which classical balls may roll and which is 'so constructed that if we let a number of little balls roll down, they will (ideally) form a normal distribution curve'[20] at the foot of the board. The normal distribution curve represents a propensity of the whole arrangement. Tilt or kick the board or take out a pin and you change the curve and the propensities 'whether or not the ball actually comes near the place from which we removed the pin'.[21]

According to Popper, this is just like the two-slit experiment. What the propensity interpretation reveals is that the behaviour of the individual system is a matter of the whole arrangement. Similarly the 'collapse of the wave-packet' is no more than a change in the specification of the experiment.

> We may say that every time the ball actually hits a certain pin (or, say, passes on its left side), the *objective probability* distribution (the propensity distribution) is 'suddenly' changed, whether or not anybody takes note of the course of the ball. But this is merely a loose way of saying the following: *if we replace the specification of our experiment by another one* which specifies that the ball hits that particular pin (or passes on its left), then we have a different experiment and accordingly get a different probability distribution.[22]

When we know which slit each photon goes through then we change the pattern on the screen.

So what is wrong with all this?

Feyerabend[23] has pointed out one considerable irony. Popper is out to attack Bohr and Copenhagenism, but experimental arrangements are exactly Bohr's *phenomena*. Both Bohr and Popper agree that probability is relational, a matter of a relation of a system to its experimental surroundings. How then do Bohr and Popper differ?

In the case of the two-slit experiment Popper manages to give a qualitative explanation of the fact that there is a difference between the sum of the two single-slit patterns and the two-slit pattern. He thinks individual quantum systems are particles with trajectories and so exploits Duane's mechanical explanation of the two-slit experiment. However there is every reason for thinking that this explanation will not do. As we noted before, you can get the two-slit effect using lasers and electron microscopes, devices that do not promise an easy application of Duane's mechanical ideas. The problem is, as Popper recognizes, not merely to avoid the paradox of the two-slit experiment but also to give an explanation of the observed two-slit pattern. There is no reason to think that Popper can do this, given his particle view of quantum systems, because Popper has no way of handling the wave aspects of quantum systems if a Duane-type explanation for interference in general fails, as it does.

If Popper fails in this, and he does, how is it that Bohr, with his somewhat similar holism, succeeds? The answer is, as Feyerabend elegantly puts it, that complementarity asserts the relational character not only of probability, but of all dynamical magnitudes. In short, Bohr is not a Popperian realist. Propensity is thus a part of, and so less than, the larger idea of complementarity.

Bohr's account of complementarity tells us that position and momentum cannot be simultaneously defined with sufficient exactitude so that a quantum system may be said to have a trajectory. This undercuts both Heisenberg's remark that our view about the past trajectory of a quantum system is a matter of taste, and Popper's claim that we may measure exact position and momentum. Bohr's response is that a precise measurement of position prevents momentum being well-defined.

Where does this leave us? We began with the puzzle of wave–particle duality, a puzzle which is embodied in paradoxical experiments, like that of the two-slit experiment, and which is expressed in quantum mechanics via the superposition principle on the one hand and by the collapse of the wave-packet on the other.

Naive resolutions of wave–particle, those which collapse the duality in favour of either wave, like Schrödinger's, or particle, like Born's, were found wanting.

Ensemble interpretations, motivated by a view about the meaning of probability statements, are often in practice articulated versions of Born's particle interpretation. Such was the case with Ballentine's version of the ensemble interpretation. Popper's, on the other hand, employed a metaphysical reinterpretation of probability in quantum mechanics. Neither succeeds. In the words of the physicist L. D. Mermin, Popper's philosophy of quantum mechanics is rubbish of a stimulating kind . . . [with which physicists should] test their own grasp of its [quantum mechanics'] foundations against Popper's view that what is most marvellously intricate and subtle in the behavior of the atomic world is just a mystery and horror to be dispelled by some clear thinking about probability.[24]

The differing metaphysics of Heisenberg and Bohr offer a powerful alternative. Both, one way or another, restrict the application of quantum mechanics.

Heisenberg employs the apparently positivistic device of proscribing some statements about quantum systems as 'meaningless'. All references to 'meaning' invoke a knee-jerk response from philosophers who jealously guard their territory. However, there is less to Heisenberg's positivism than meets the eye. In his view, it is quantum theory, not a theory-independent account of perception or evidence, which decides what is meaningful in quantum theory.

Bohr has a deeper and more metaphysical interpretation of quantum theory. There are limitations on what we can describe and explain. The subatomic world is an abstraction. We are restricted to describing wholes which incorporate that abstraction, along with our attempts to couple with it. The extent of this unavoidable restriction on our capacity to describe and explain is expressed in the indeterminacy relations.

Against Bohr stands Einstein, arguing for the incompleteness of quantum theory and, implicitly, for its supplementation by a deeper theory of the individual quantum system. The view that quantum theory is as far as *we* can go – there may be more to the world but *we* can't get at it – is apparently beset by paradox, including the troublesome feature of nonlocality. The metaphysics of complementarity is a 'tranquilizing philosophy' which explains nothing.

So can quantum theory be supplemented by a deeper, more classically intelligible theory or interpretation? Can such a theory or interpretation avoid the paradoxes? Most interesting of all, if it cannot avoid the paradoxes, can altering our logic to a logic that drops out so nicely from quantum theory itself, quantum logic, help us?

These are the questions that occupy us in Part II.

Part II

6

Quantum mechanics for natural philosophers (II)

Quantum mechanics comes in a variety of different forms and formalisms. There is wave mechanics, matrix mechanics, Dirac's version of quantum mechanics, von Neumann's version of quantum mechanics. The philosopher of physics has to ask not only what quantum mechanics means, but also what it is.

Orthodoxy in the philosophy of physics treats quantum mechanics as the general theory of microphysics which takes *Hilbert space* as its state-space and which associates observables with certain special operators on that space. This amounts to a decision as to what quantum mechanics is. However it is a far from arbitrary decision. For the Hilbert space, or von Neumann, formalism summarizes and captures everything that is in the other versions of quantum mechanics, without any of the dubious mathematics of, for example, Dirac's version.

Contemporary philosophers of physics take a very formal view of what quantum mechanics is, a view which has many advantages and some disadvantages. Whatever the disadvantages of focussing on the formalism of quantum mechanics – and it may lead to losing sight of the 'physics' and possibly to a bias against Bohrian Copenhagenism – it is certainly true that the literate philosopher of physics must have some acquaintance with Hilbert space.

There are in fact several topics we have to deal with, though we deal with them only very cursorily. (This chapter is no substitute for Jauch's classic text *Foundations of Quantum Mechanics*.) Apart from Hilbert space itself, we need a handle on the general quantum-mechanical notions of state and observable. We need to understand the most general notion of quantum-mechanical state – the mixture – and the best way of describing it, via the density operator. We need to know what a lattice is, because quantum logic is a lattice in the first instance. We need to know why quantum logic is non-Boolean. We also need to know how to combine the quantal descriptions of several quantum systems.

First we ask: What are the advantages and disadvantages of the Hilbert space view of quantum mechanics? First, the advantages. Some of the no-hidden-variable proofs crucially depend on taking Hilbert space as the state-space of quantum physics. Certainly some no-hidden-variable theorems do. Second, quantum logic is a structure that drops out of Hilbert space. Thus the discussion of two of the most interesting issues in the philosophy of quantum mechanics – the possibility of hidden-variable theories and the role of quantum logic – depends directly on the structure of Hilbert space. This is not true of *all* such issues. The violation of Bell's inequality in quantum mechanics, for example, and the question of nonlocality depend on the treatment of a particular correlated system, like the singlet state of a pair of correlated electrons without very much in the way of commitment to the Hilbert space model.

One might of course *stipulate,* as some philosophers would if pressed, that quantum mechanics *is* just applied Hilbert space theory. Discussing quantum mechanics without reference to the Hilbert space model would then be rather unattractive. But such a stipulation would involve ignoring both the history of the subject, to which the Hilbert space unification was a relatively late development, and also the fact that there are alternative approaches to formalizing quantum mechanics which are currently under discussion.

One possible philosophical disadvantage of too formal a view of quantum mechanics has been pointed out by Feyerabend.[1] As one's treatment of any physical theory, but of quantum mechanics in particular, becomes less and less formally *im*perfect, the empirical meaning of the theory tends to become more and more obscure. The 'physics' tends to get lost in the mathematics, a fact which can be philosophically significant. Thus, if Bohr has no 'problem of measurement' at least in the way that von Neumann does, it is precisely because Bohr does not treat measurement as a problem in the way that von Neumann does. Bohr is more interested in the 'physics' than in the formalism. This brings out yet again an interesting but neglected problem in the philosophy of physics: whether one really can distinguish between 'the physics' and 'the formalism', particularly for a theory like quantum mechanics which defies physical intuition.

Whether Bohr's disdain for the quantum-mechanical formalism is or is not an advantage is a large topic. We however take the conventional view that the philosophy of quantum mechanics should be conducted via the Hilbert space formalism. This has the advantage for us that it will take us naturally on to quantum logic, which is very difficult to motivate without the Hilbert space model.

What is Hilbert space?

The ordinary three-dimensional physical space in which we live can be represented by a set of vectors each of which can be expressed as a sum of three *basis* vectors **i, j,** and **k** chosen in the x-, y-, and z-directions. The space is three-dimensional because three vectors suffice for a basis: any vector in the space can be represented as a weighted sum of the three basis vectors. In this space, two vectors are orthogonal if and only if they are at right-angles. We also have the product of two vectors. This is the number (it isn't a vector in the space) which is equal to the product of the sizes of the two vectors and the cosine of the angle between them.

The space of vectors which represent positions in physical space is *real* in that you can multiply a vector by a *real* number and produce another vector, in the same direction, lengthened or shortened depending on whether the number is greater or less than 1. Therefore one says that ordinary physical space is a vector space over the field of the reals, for multiplying a vector in our space by a complex number makes no sense (or rather at best it produces an object which is not a member of our space).

Compare this three-dimensional space with the space of the quantum mechanical *states* of a quantum system.

State-vectors can be added, as their name suggests. This much follows from the superposition principle. Just as important is the fact that any state-vector can be represented as a sum, as it happens at most a countable sum, of other state-vectors. We can represent any state-vector as a sum of suitably chosen *orthogonal* state-vectors. Two state-vectors f and g are orthogonal if and only if their inner product

$$\langle f|g\rangle = 0.$$

Orthogonality is clearly a symmetrical but not necessarily transitive relation. Orthogonality is a kind of 'being at right-angles', and the inner product $\langle f|g\rangle$ is a generalization of the scalar product of vector in our real 3-space. Unlike that scalar product, the inner product of two state-vectors need not be (and usually isn't) a real number.

The Hilbert spaces of quantum mechanics are vector spaces just like our real physical space except first that they are not restricted to being three-dimensional, second that you can multiply their vectors by *complex* numbers to produce another member of the space, third that the vectors in them are all of unit length (except the zero vector), and fourth that their elements are interpreted as being associated with the states of quantum systems rather than with locations in a physical space.

The elements are state-vectors, that is wave-functions, more or less. In fact, they are all the square-integrable functions (wave-functions plus some more) whose square-integrals equal 1, and, equivalently, they are column matrices the squares of whose (complex-valued) elements add up to 1. The business of 'equalling 1' is called *normalization.* We came across it in Chapter 2. The fact that there are these two very different realizations of Hilbert space is really the fact that wave mechanics and matrix mechanics are observationally equivalent, something Schrödinger demonstrated in 1926.

A Hilbert space naturally has enough structure to have a dimension, an idea which is defined as follows. Consider a state-vector which cannot be built out of the elements of some set of state-vectors by adding (some or all of) them together in some proportions. (The proportions will be complex numbers in general.) The original state-vector is said to be *linearly independent* of the set. One can have a set of state-vectors which is such that every member of it is linearly independent of the rest. The size of the largest set of mutually linearly independent state-vectors is then the dimension of the Hilbert space.

The two interesting points in all this are that quantum-mechanical Hilbert spaces are vector spaces over the field of complex numbers and that their dimension may be any integer from 1 to a countable infinity. (For completeness one should say that infinite dimensionality generates topological subtleties not apparent in our ordinary 3-space which by and large are not of direct concern to us.)

Recall that the wave-function, or state-vector of an electron (say) can be a function $\psi(x,y,z)$, where x, y, and z are some chosen axes of our real physical 3-space. The function $\psi(x,y,z)$ is an element of the Hilbert space of states of the electron. Of course, it is important not to confuse the Hilbert space of states with the real physical 3-space which supplies the variables x, y, and z. The Hilbert space is a highly abstract object – a space of states. That is, it is just the set of all possible states of a quantum system but it has a structure, due to the superposition principle, which makes it a bit like the structure of the space we live in.

In some ways some Hilbert spaces are very *unlike* real 3-space. The space of states of an electron is *infinite dimensional.* But a bound is placed on the size of this infinity. It is always assumed that a Hilbert space can have no more than a countable infinity of linearly independent state-vectors. This implies that there are no eigenstates of exact position, that the Dirac δ-function is illegitimate.

Imagine a quantum system, an electron say, and consider the set of state-vectors which give information about position and momentum. Forget about spin for the time being. Clearly there is a nondenumerable infinity of possible electron states and therefore a nondenumerable infinity of state-vectors. It follows, or very nearly follows, that there can be no state-vectors for exact position. If there were there would be a nondenumerable infinity of them since otherwise they would not be linearly independent, and this they surely would have to be. There can however be state-vectors which vanish everywhere except for values of position x, y, and z in nonzero intervals as small as you like. Exactly similar remarks go for momentum.

For our purposes the commonest Hilbert space in quantum mechanics, the space of state-vectors for position and momentum, has *too much* structure. It is too large. The Hilbert space of intrinsic spin states of an electron, which has only two dimensions and is therefore the simplest nontrivial example, is complex enough to illustrate all the features of Hilbert space we need. The one exception concerns the *orthomodularity* of quantum logic, a feature which depends on the infinite dimensionality of the quantum-mechanical Hilbert space. Though this two-dimensional Hilbert space is over the field of the complex numbers one can draw it, almost well enough, and one can see it, very nearly, with one's mind's eye as an ordinary real vector space, like the space of vectors on a piece of paper.

States and observables

Compared with classical mechanics, quantum mechanics employs a very indirect notion of the state of a system. It also contains an altered treatment of observables.

In quantum mechanics observables are associated with *Hermitian* operators on Hilbert space, this to ensure that their eigenvalues – their possible determinate value – are real rather than complex. For any operator A one can define its Hermitian conjugate $A\dagger$ by requiring that for any state-vectors $|f\rangle$ and $|g\rangle$

$$\langle g|A\dagger|f\rangle = \langle f|A|g\rangle.$$

An operator A is Hermitian if and only if it is equal to its Hermitian conjugate. The eigenvalues of any observable are expected to be real to ensure that their ordering is linear, and is the same as the possible results of experiment.

It is natural to associate observables with a related class of operators called *projection operators*. A projection operator P is idempotent, which is to say 'once is enough'. For any state-vector $|f\rangle$

$$P(P|f\rangle)=P|f\rangle.$$

Projection operators reflect the idealness of ideal measurements: repeated measurement yields the same result. For example, in a Stern–Gerlach experiment set in a given direction there is a projection operator associated with spin 'up' and one associated with spin 'down'. The effect of the projection operator associated with spin 'up' is to project any state-vector into a state in which spin 'up' is definite. There are only two definite values for spin 'up': 'yes' and 'no'. The effect of making a measurement of spin is to project the old state-vector onto one of definite spin 'up' or 'down', definite in that repeated measurement results in no more projecting. The subject of projection operators leads on naturally to the subject of subspaces of Hilbert space.

A subspace of a Hilbert space is a set of state-vectors closed under addition. If $|f\rangle$ and $|g\rangle$ are members of a subspace S then so is

$$c_1|f\rangle+c_2|g\rangle.$$

Subspaces are essentially sets of state-vectors but not just any old sets: they are sets 'closed under superposition'. Drawing on our physical 3-space analogy subspaces are either fixed directions in space or fixed two-dimensional planes (or, trivially, the zero vector or the whole 3-space). The first important point is that subspaces and projection operators are in one-to-one correspondence. A subspace is the set of state-vectors into which some projection operator projects any state-vector. The second important point is that subspaces (and hence projection operators) are in a natural one-to-one correspondence with 'propositions true of the associated quantum system'. This leads us on to quantum logic which, to begin with, is nothing more than the lattice of closed subspaces of a Hilbert space.

What is quantum logic?

Quantum logic is first of all a special sort of lattice.

A lattice is a set with a *partial ordering relation* and a pair of operations – a *meet* and a *join* – which are closed on the set. If the set is a set of propositions the properties of the meet and the join correspond to the expected basic properties of conjunction and disjunction respectively.

A partial ordering relation is reflexive, antisymmetric and transitive.

Therefore set-theoretic inclusion is a partial ordering relation on a collection of sets. The subspaces of a Hilbert space associated with a quantum system correspond to propositions about that system and are just special subsets of the overall domain of the Hilbert space. In the lattice which is quantum logic the elements are the subspaces of such a Hilbert space, the partial ordering relation is set-theoretic inclusion defined on the set of subspaces. The operation of meet is set-theoretic intersection but the operation of join is not set-theoretic union but span for reasons which depend on the superposition principle. The span of two subspaces is always at least as large as their union. When one transcribes this lattice as a logic, (when you represent it as a logic rather than as an algebraic structure) the inclusion relation corresponds to 'implication', the meet to 'and' and the join to 'or'.

The most important deficiency of quantum logic is that it is *nondistributive,* a fact which can be illustrated by considering a two-dimensional real Euclidean space which is just a very simple example of a Hilbert space. First we should consider what language quantum logic is the logic of.

When one says that quantum logic is the logic of the language of quantum mechanics, the language one is referring to is not the mathematical language of the formalism but is something special, 'the elementary language of quantum mechanics',[2] hereafter ELQM.

The basic sentences of ELQM have the form

At time t the value of the dynamical variable d for the system S lies in the range D.

When we get on to axioms, rules of inference and to semantics it will turn out that quantum logic is nondistributive. This is the single most important fact about it. But quantum logic and classical logic share many features. We shall interpret quantum logic bivalently in the sense that every proposition will be either *true* or *false*. Of course unlike classical logic quantum logic cannot be truth functional. As a corollary, the quantum logical connectives cannot be defined by means of truth tables. So there arises a philosophical problem about *the meanings of the quantum logical connectives.*

Before we deal with quantum logic proper we need some motivation, and we need to examine the lattice which is quantum logic. Then we return to quantum logic as logic.

Why the distributive law is sometimes false

For a given system there is, corresponding to a given range D of a dynamical variable **d,** a projection operator $P^{\mathbf{d}}{}_D$. There is a set of vectors of the Hilbert space S which are unaffected by the action of $P^{\mathbf{d}}{}_{D}$. These vectors which make up a subspace of the Hilbert space $P^{\mathbf{d}}{}_{D}$ will disturb the other state-vectors of S and project them into this subspace. The projection postulate, in all its versions, gives good reason to associate the sentence of ELQM which asserts that a quantum system has **d** in D with this subspace: that proposition will be 'true' for any state-vector in the subspace and 'false' for all others.

This tells a reasonable story about the basic sentences of ELQM. But what about the compound sentences?

The set-theoretic *intersection* of two subspaces is another subspace. So the set of states for which both of two sentences of ELQM are true can be defined to be the intersection of the two corresponding subspaces.

However, the union of two subspaces is not necessarily a subspace. What we must associate with the disjunction of two sentences of ELQM is the smallest subspace which set-theoretically includes both the subspaces corresponding to the disjuncts. This subspace is called the *span* of the two subspaces and includes their union. For finite-dimensional subspaces it is the subspace consisting of all vectors which can be made out of the vectors in the union of the subspaces. This includes the two subspaces, but is larger than their set-theoretic union. We restrict this remark to the finite-dimensional subspaces because the infinite-dimensional case is even more complicated and we consider it later. But suffice it to say that so far '&' in quantum logic looks normal and 'v' looks decidedly peculiar.

So why does the distributive law sometimes fail?

The simplest and indeed the stock example is given by electron spin: the case of the two-dimensional Hilbert space of a spin-$\frac{1}{2}$ system.

We know that any state-vector in the Hilbert space may be represented as a superposition, a weighted sum (the weights are *complex* numbers c_i which are such that $\Sigma_i |c_i|^2 = 1$) of the 'up' and 'down' spin states for any chosen direction. So choose the z-direction for convenience. Suppose that we have an electron whose spin is 'up' not in the z-direction but in the x-direction.

Let a be the sentence of ELQM that asserts (truly as it happens) that the electron has spin 'up' in the x-direction. And let b and c be the sentences of ELQM which assert that S has spin 'up' and 'down' respectively in the z-direction. The disjunction $b \vee c$ corresponds to the whole subspace and so is true, and the subspace corresponding to a & $(b \vee c)$ is the subspace corresponding to a. In lattice-theoretic terms

$$a \wedge (b \vee c) = a.$$

But the subspace corresponding to both a & b and a & c is empty and so is the subspace corresponding to their disjunction. Therefore

$$a \wedge (b \vee c) \neq (a \wedge b) \vee (a \wedge c).$$

Quite generally in quantum logic the right-hand side of this expression is 'logically stronger' and always implies the left-hand side though not always conversely. When this converse fails so does distributivity.

One can set up 'an elementary language of classical mechanics' and one should contrast this case with the quantum-mechanical one. In 'ELCM' distributivity will reign. Why? Because our state space is a phase space and not a Hilbert space and we associate propositions with *all* sub*sets* of elements of phase space and not just special subsets (subspaces). A disjunction in ELCM corresponds to a set-theoretic union of subsets and so, like the lattice of subsets of a set, ELCM is distributive. The logic of classical physics is, quite generally, classical.

Therefore we say that the logic of classical mechanics is classical and that the logic of quantum mechanics is nonclassical. However, this remark is limited to the elementary languages of the two theories. There is nothing nonclassical about the logic of the formalism of quantum mechanics. What is nonclassical is the totality of the logical relations in which statements of fact about quantum systems stand to one another. Therefore the radically metaphysical thought that quantum mechanics shows that the 'logic of the world' is nonclassical must accommodate the fact of classical mathematics in the formalism of quantum mechanics.

Quantum logic as a lattice

As well as the binary operations of *meet* and *join* the lattice which is quantum logic has a third primitive operation, the unary operation of *orthocomplementation*. Meet and join correspond to 'and' and 'or'. Orthocomplementation corresponds to 'not'. Quantum logic is an orthocomplemented lattice with a bit of extra structure but not enough extra structure to make it classical, that is, a Boolean algebra.

Formally, quantum logic is the lattice Q whose base set is L such that

$$Q = \langle L, \mathbf{1}, \mathbf{0}, \leqslant, \wedge, \vee \perp \rangle$$

which satisfies

(1) $a \wedge b \leqslant a$
(2) $a \wedge b \leqslant b$
(3) if $c \leqslant a$ and $c \leqslant b$ then $c \leqslant a \wedge b$

and their duals

(4) $a \leqslant a \vee b$
(5) $b \leqslant a \vee b$
(6) if $a \leqslant c$ and $b \leqslant c$ then $a \vee b \leqslant c$.

The conditions (1) to (3) and (4) to (6) are the basic properties of the join and meet respectively. Given (1) to (6) the *principle of duality* holds: for any theorem concerning a lattice you can get another by interchanging $\wedge$ and $\vee$ and $\leqslant$ and $\geqslant$ (where $a \leqslant b$ if and only if $b \geqslant a$).

What about 'negation'?

In any orthocomplemented lattice there are a maximum element **1** and a minimum element **0** which, for all a in L, are such that

(7) $a \leqslant \mathbf{1}$,
(8) $\mathbf{0} \leqslant a$.

The orthocomplementation maps each element in L to its orthocomplement

$\perp : L \rightarrow L$
$\perp : a \mapsto a^{\perp}$

such that

(9) $a \vee a^{\perp} = \mathbf{1}$
(10) $a \wedge a^{\perp} = \mathbf{0}$
(11) if $a \leqslant b$ then $b^{\perp} \leqslant a^{\perp}$
(12) $(a^{\perp})^{\perp} = a$.

A 'negation' which is an orthocomplementation satisfies the 'principle of noncontradiction', 'the law of excluded middle' and 'double negation'.

Conditions (1) to (12) capture the structure possessed by a particular kind of lattice known as an ortholattice – which is no more than an unspecial lattice with an added orthocomplementation. Quantum logic has this much structure and some more. One can capture the extra structure that quantum logic has by a further axiom known as the 'orthomodular law'. This is essentially a restriction on distributivity. The distributive law holds for more substitution instances in quantum logic than it does in the general ortholattice, but it does not hold for all substitution instances as it does classically.

The lattice of subspaces of a *finite-dimensional* Hilbert space is *modular*. It satisfies

$$a \wedge (b \vee c) = (a \wedge b) \vee (a \wedge c)$$

whenever $c \leqslant a$. Modularity represents a weakening of distributivity.

In the *infinite-dimensional* case, which is the one that is generally of interest in quantum mechanics, even modularity fails and distributivity has to be further weakened. The lattice of closed subspaces of an infinite-dimensional Hilbert space is *orthomodular*. That is, it satisfies

$$a \wedge (b \vee c) = (a \wedge b) \vee (a \wedge c)$$
$$\text{whenever both } c \leqslant a \text{ and } b \leqslant a^{\perp}.$$

Distributivity is further weakened in the infinite-dimensional case because the span of two infinite-dimensional subspaces contains 'limit vectors' which cannot be formed from *finite* sums of vectors from the two subspaces but only from the sum of an infinite series of such vectors. Otherwise the span would not be closed.

Mixtures

The ensemble interpreter thinks that state-vectors describe ensembles. To explore the important idea of mixed ensembles, or *mixtures,* we temporarily adopt his idiom.

Suppose you have two equally intense beams of electrons, the first consisting of electrons with spin 'up' in the x-direction, the second of electrons with spin 'up' in the z-direction. The two beams can be described by the state-vectors $|+x\rangle$ and $|+z\rangle$ respectively. All the electrons would be found to have spin 'up', in the x-direction in one case, and in the z-direction in the other. These original beams are intuitively 'pure' in a sense that we make precise later.

Now merge or 'mix' the two pure beams in equal proportions to form a new beam. How should we describe the resulting mixture?

One thing we can say is that we expect the new description to give the correct statistical behaviour of the mixed beam. Thus we would expect that it would tell us that three quarters of the electrons in the new beam would be found to have spin 'up' in the x- and z-directions (all of them in one subbeam and half in the other) and one quarter 'down' in those directions (none in one subbeam and half in the other). One would also expect that one half the electrons in the mixed beam would be found to have spin 'up' (or 'down') in the y-direction, as the two pure subbeams have a random orientation of spins in the y-direction.

So why not simply *superpose* the two original state-vectors and obtain a new state-vector?

The answer is that this description would give you the wrong statistics. Any superposition of $|+x\rangle$ and $|+z\rangle$ will be an eigenstate of spin 'up' in some direction simply because it will be a state-vector. All the electrons in any chosen superposition of $|+x\rangle$ and $|+z\rangle$ will be found to have spin 'up' in some direction.

For example, the superposition

$$|\rangle = 0.541|+x\rangle + 0.541|+z\rangle$$

represents not a mixture but a pure ensemble all of whose electrons have spin 'up' in the xz-plane at 45° to the $+x$- and $+z$-axes. (Note that $|+z\rangle$ and $|+x\rangle$ are not orthogonal so that the sum of the squares of their peculiar coefficients is not equal to 1.)

But if an ensemble is a *mixture* of two pure subensembles of spin 'up' in the x- and y-directions, there will be *no* direction in which all the electrons have spin 'up'.

There is an extended but simple formalism for handling the description of mixed beams – the density operator formalism. The density operator provides a way of 'mixing' the descriptions of pure ensembles to produce something which is not a pure ensemble but is a genuine 'mixture'. The density operator formalism incorporates the state-vector formalism as a special case. It is usual nowadays, among philosophers even more than among physicists, to formulate the quantum algorithm entirely in the more general density operator formalism even when only state-vectors are considered. We shall find that we need density operators, rather than state-vectors, when we discuss the problem of measurement in quantum mechanics in Chapter 7.

The density operator

Consider this object

$$|+x\rangle\langle+x|.$$

Unlike $\langle+x|+x\rangle$, which is a complex number, namely the scalar product of $|+x\rangle$ with its complex conjugate $\langle+x|$, $|+x\rangle\langle+x|$ is an *operator*. If you multiply it into a state-vector like $|+y\rangle$ you get

$$|+x\rangle(\langle+x|+y\rangle)$$

which is $|+x\rangle$, a state-vector, multiplied by the complex number inside the brackets.

One can, rather trivially, represent the state-vector $|+x\rangle$ by the operator – called the *density* operator for reasons we note later –

$$\rho = |+x\rangle\langle+x|.$$

The operator ρ operates on $|+x\rangle$ to produce $|+x\rangle$ as

$$(|+\rangle\langle + x|)|+ x\rangle = |+ x\rangle(\langle + x| + x\rangle.$$

It is easy to see that ρ operates on any other state-vector$|\ \rangle$ to produce$|+ x\rangle$again but this time shrunk by a factor, the scalar product

$$\langle + x|\ \rangle.$$

But$|+\rangle\langle + x|$is a trivial example and very much a special case. In general the object

$$\rho = \Sigma_i\lambda_i|i\rangle\langle i|$$

where $\Sigma_i\lambda_i = 1$, and the$|i\rangle$ are normalized state-vectors, is a *density operator* and it can be used to describe a genuine mixture. It is unlike the superposition description in that it gives the correct statistics for a mixture composed of pure subensembles each of which is described by the $|i\rangle$ each with weighting λ_i. The λs give the 'densities' of the sub-ensembles. Furthermore the expression appears in classical statistical mechanics where it is called 'the density operator'. Hence the use of the term in quantum mechanics.

For now we consider density operators in which the $|i\rangle$ are orthonormal or, more accurately, in which $\langle i|j\rangle = 1$ if $i = j$ and $= 0$ otherwise. In this case it is easy to see that the λ_i are the *eigenvalues* of the operator and the corresponding *eigenstates* are $|i\rangle$.

Density operators are operators and so you can multiply them. If you multiply a density operator ρ by itself to get ρ^2, then it is easy to see that

$$\rho^2 = \Sigma_i\lambda_i^2|i\rangle\langle i|$$

If $\Sigma_i\lambda_i = 1$, then $\Sigma_i\lambda_i^2\langle 1$, unless all the λs except for one are zero. In this latter case the mixture is really a pure case.

So, here is a general property of density operators ρ.

$$0\langle\rho^2\leqslant\rho$$

and if $\rho^2 = 1$ then ρ describes a *pure* ensemble.

This mathematical fact can be used as a definition of 'pure case' or 'pure ensemble' which conforms to our intuition that an ensemble described by a single state-vector is homogeneous. When $\rho^2\langle\rho$, then ρ describes a *proper mixture*. A pure ensemble is then an *improper mixture*.

How do density operators give the right statistics for mixtures? There is a neat algorithm for getting the statistics from a density operator.

First, we need the idea of the *trace* of an operator.

The trace of an operator is the sum of its diagonal elements. We can find the matrix of an operator **O** in any representation (say) $|k\rangle$, and the value of the trace

$$\text{trace}(\mathbf{O}) = \Sigma_i \langle i|\mathbf{O}|i\rangle$$

will be fixed. It is the same for any choice of $|i\rangle$. This is a remarkable but very simple property of the trace.

It is not too difficult to see that we can formulate the expectation value for an observable **O** for an ensemble (pure or mixed) described by the density operator ρ as

$$\langle \mathbf{O} \rangle = \text{trace}(\rho \mathbf{O}).$$

There is one *calculation* we need to make, even as philosophers. In discussing the quantum-mechanical violation of Bell's inequalities we need to know the probability of getting the result 'up' for a spin measurement made on an electron at angle θ to the vertical, z-axis, given that the electron has its spin 'up' in the z-direction.

We can take it that the direction is in the xz plane without any loss of generality, because the final answer makes no mention of x. For ask: what is the operator o_θ for a spin measurement at an angle θ to the vertical? We represent the answer using the Pauli spin matrices. Clearly

$$o_\theta = o_z \cos\theta + o_x \sin\theta$$

$$= \tfrac{1}{2}\hbar \begin{pmatrix} \cos\theta & \sin\theta \\ \sin\theta & -\cos\theta \end{pmatrix}.$$

It is easy to check that the state-vectors for sign 'up' and spin 'down' in the direction θ to the z-axis are

$$\begin{pmatrix} \cos\frac{1}{2}\theta \\ \sin\frac{1}{2}\theta \end{pmatrix} \quad \text{and} \quad \begin{pmatrix} -\sin\frac{1}{2}\theta \\ \cos\frac{1}{2}\theta \end{pmatrix}.$$

Given that the electron has its spin 'up' in the z-direction its state-vector is

$$\begin{pmatrix} 1 \\ 0 \end{pmatrix}.$$

If this is a superposition of the two vectors for spin 'up' and 'down' in the θ-direction, then

$$\begin{pmatrix} 1 \\ 0 \end{pmatrix} = c_{\text{'up'}} \begin{pmatrix} \cos\frac{1}{2}\theta \\ \sin\frac{1}{2}\theta \end{pmatrix} + c_{\text{'down'}} \begin{pmatrix} -\sin\frac{1}{2}\theta \\ \cos\frac{1}{2}\theta \end{pmatrix}$$

giving $c_{\text{'up'}} = \cos\frac{1}{2}\theta$ and the probability $\cos^2\frac{1}{2}\theta$, which is the answer we want.

One last question: How should we combine two Hilbert spaces?

The problem arises in two ways. First, for a single quantum system like an electron one can combine the Hilbert space of its state-vectors which are functions of position with the Hilbert space of state-vectors for its intrinsic spin. Second and less trivially, one may want to describe, and hence combine the Hilbert spaces of, two different but interacting quantum systems each with its own Hilbert space. This second case is one that arises when one tries to give a quantum-mechanical account of the measuring process, treating the measuring system as a second quantum-mechanical system interacting with the measured system.

Let H_1 and H_2 be two Hilbert spaces, associated with the systems I and II, and let f_1 and f_2 be state-vectors belonging to the H_1 and H_2 respectively. The tensor product $H_1 \otimes H_2$ of the two Hilbert spaces is spanned by the 'product' vectors $f_1 f_2$. If f_1 and f_2 are given concrete representations (say as wave-functions) then their product is obtained by simply multiplying them.

What is the point of all this? First note that the dimension of the tensor product space is the product of the dimensions of the two Hilbert spaces. But if the systems I and II are *correlated,* so that if I is in the state $f_{1,\mathrm{I}}$ and II is in the state $f_{2,\mathrm{II}}$ then the state of the combined system is a superposition of terms like

$$f_{1,\mathrm{II}} f_{2,\mathrm{II}}$$

with all the 'cross' terms having zero coefficients.

This is the case that arises in EPR and also reappears when we discuss von Neumann's theory of quantum-mechanical measurement.

7

Projection postulates

The concept of measurement plays both a central and a problematic role in quantum mechanics.

First of all, and unusually, the word ‘measurement’ figures in the fundamental axioms of quantum mechanics – in principle (3) of Chapter 2 for example – and in this quantum physics is quite unlike classical physics as a whole. For though a measurement made on a classical system naturally involves an interaction between the system and a measuring apparatus, there is nothing special according to classical physics about measurement interactions. Describing the measuring process is a straightforward problem of applied classical physics. The classical laws of motion do not break down during the measuring process.

For example, suppose you observe a classical particle using light bounced off it. In the classical account you can make the disturbance on the observed system as small as you like and, ‘in principle’, you can observe the system with arbitrary accuracy. ‘In principle’ – a favorite expression of the philosopher of physics – means what it always means in physics. The contrast is with ‘in practice’. Of course, in practice there are all sorts of limitations on measurement. There are also some theoretical limitations, like those due to thermodynamics. But these are not the limitations imposed by the theory we are considering, namely classical mechanics. Such limitations as there are are imposed by some other theory. (The expression ‘in principle’ tends to be used when we are discussing the principles of a particular theory and, theoretically but not in practice, physics is a collection of disjoint theories.)

Measurement in quantum mechanics is quite different from measurement in classical physics. One can easily beg questions in the philosophy of quantum mechanics if one says exactly how it is different. But in the orthodox view, the Schrödinger equation for a quantum system generally *breaks down* during the measuring process. Describing a quantum-mechanical measurement is not a straightforward problem of applying the Schrödinger equation.

In Heisenberg's γ-ray microscope thought-experiment we cannot observe the electron with arbitrary accuracy because the photon we bounce off it has a fixed momentum for a given wavelength of light, and an indeterminate amount of this momentum is transferred to the observed electron. But Heisenberg's thought-experiment is semiclassical.

In a fully quantum-mechanical treatment we would have to violate Schrödinger's equation at some point. An account of measurement which fully respected Schrödinger's equation would tell us that the macroscopic measuring instrument would not finish up in a determinate state but would rather be in something akin to a superposition of states. However, we know that measurements always leave the macroscopic measuring apparatus in a determinate state.

In addition to the fact that the concept of measurement appears in the fundamental principles of quantum mechanics, measurement in quantum mechanics is mysterious.

For example, suppose you take a stream of electrons and measure their z-spin with a Stern–Gerlach apparatus. Let us say that half have z-spin 'up' and half their z-spin 'down.' Take the stream that now has z-spin 'up' and remeasure z-spin and you will get that all the electrons still have spin 'up' (ditto for the spin 'down' ensemble). Nothing paradoxical here.

Now take the two streams of z-spin 'up' and 'down' and separately measure 'x-spin' for each, and you find that half of each stream gives x-spin 'up' and half 'down'. Nothing paradoxical here either.

Make a *record* of which stream each electron belongs to and then remix the two streams of electrons of x-spin 'up' and x-spin 'down' from the z-spin 'up' stream. Then remeasure z-spin, and you will find that the remixed stream now has randomized z-spin, half of the electrons have z-spin 'up' and half have z-spin 'down.' Again nothing paradoxical, unless disturbance by measurement is itself paradoxical. We simply say that the *recording* of each electron's x-spin *disturbed* it and interfered with its z-spin.

Alternatively, remix the two streams of electrons with x-spin 'up' and x-spin 'down' *without* 'making a record of which stream each electron belongs to'. Remix the streams *without disturbance,* and you find the re*superposed* streams are of z-spin 'up' electrons. This really is paradoxical. It is as if, in the previous example, *looking* at the streams flipped them into definite x-spin streams. If you do not *look* at them, that is if you do not 'make a record', you can *superpose* them as opposed to merely mixing them.

To put it another way, a measurement interaction creates mixtures, whereas a nonmeasurement interaction does not. Measurements generally violate the Schrödinger equation, nonmeasurement interactions obey it. That is very strange.

Most, perhaps all, of the paradoxes of quantum mechanics involve measurement at some point, because the paradox, whatever it is, becomes apparent only when we get some results, a 'record'. Think of the two-slit experiment with the *measurement* of the photon's location on the screen, or EPR with the effect on particle II of *measurements* on particle I. In fact one can claim with only a little exaggeration that the problem of measurement is *the* central problem of the philosophy of quantum mechanics. So what does quantum mechanics have to say about the measurement process? A minimal, and uncontroversial, partial answer is the following.

Suppose a quantum system S is in the state $|\,\rangle$. Represent the state-vector $|\,\rangle$ in the basis $|a_i\rangle$ which are the eigenstates of the operator A corresponding to the observable **A,** so that

$$|\,\rangle = c_i|a_i\rangle$$

where $\Sigma_i|c_i|^2 = 1$.

The probability interpretation of the state-vector tells us that, idealizing a bit, a measurement of **A** will give one of the values a_i each with probability $|c_i|^2$. You will get the result that the value of **A** is a_k with certainty, which is with probability equal to 1, if and only if

$$|\,\rangle = |a_k\rangle.$$

In other words, it follows from the probability interpretation of the state-vector that a measurement of **A** on the system S will yield a_k *with certainty* when the state-vector of the system is the eigenstate $|a_k\rangle$. This much is generally agreed. It is the residue of Born's original interpretation of the wave-function. But there is much more to be said as we shall see.

One also has to ask:

What is a measurement?

First of all, if one aspect of a measurement is the production of some permanent macroscopic record of the result, and one would expect that it would be, then the interaction will involve a macroscopic apparatus at some point. One would expect that any measuring process would terminate in the irreversible fixing of some macroscopic measuring apparatus and that this should contain a *record* of the value of some observable.

Second, a measurement of the **A**-value of a system S will necessarily involve some physical interaction with S, although this interaction need not *disturb* S, need not throw S into a *new* state, even though it can be expected to throw the measuring apparatus into a new state.

In what we might call an **A**-measurement we set up some measuring apparatus which interacts with the measured system in such a way that the **A**-value of the system S is registered on the apparatus, or, less directly, in such a way that the **A**-value of S can be inferred from the behaviour of the apparatus.

However not *all* interactions between a quantum system and a macroscopic apparatus count as measurements. So what sort of interactions are measurement interactions?

Imagine a measuring apparatus which, like a Stern–Gerlach apparatus, does not destroy the system which it measures. Ideally, measurements made on a given system with the apparatus can be repeated immediately. Now suppose that a 'measuring device' yields the result that the **A**-value of a system S is (say) a_1 but, when immediately used to repeat the 'measurement', it yields a different result (say) a_2, and so on for further 'measurements'. In what sense is our somewhat randomizing 'measuring device' really making measurements? One of the values a_1, $a_2, \cdots, a_n$ may be the right value. But it may be that none is. Then the 'measuring device' is no more making measurements than a random sequence of photographs of a clock (the photographs taken at different times) tells the time.

If an interaction between a system and a measuring device is a measurement just in cases where it produces the same result when repeated immediately, then there may be no such things as measurements. But luckily for us, there are. We can take as an empirical fact that to a good approximation there are such measurements. They are called, in Pauli's terminology, *measurements of the first kind,* or *ideal measurements.* Measurements with Stern–Gerlach apparatuses are good examples of the type. The essential properties of ideal measurements are that they are repeatable and give the same result when immediately repeated – that is, as the time between the measurements goes to zero, so does the difference between the two results.

The *idempotency* of ideal measurements is not a necessary feature of all measurements. In other words, there are nonideal measurements. Idempotency cannot be a feature of measurements which destroy the system measured. An example of such a measurement would be the measurement of a photon's energy by absorbing it and measuring the momentum it transfers to the absorber. One can have good empirical or theoretical grounds for thinking that such a measuring apparatus does measure, without necessarily being able to repeat the measurement on a given system. These are measurements whose reliability rests on the our faith in other measurements. However *ultimately* our experimental data must presuppose the existence of some measurements of the first kind.

It follows from the probability interpretation of the state-vector that measurements of the first kind project the state-vector of the measured system onto an eigenstate of the measured observable. This idea is called the projection postulate, in its most restricted form. There is room for many different extensions of, or versions of, the projection postulate each of which satisfies this requirement and the choice between them is empirical. We want a projection postulate which can handle measurements on mixtures and, even more importantly, a projection postulate which can handle *nonmaximal* measurements.

Maximal and nonmaximal measurements

The projection postulate, as characterized so far, tells us that a measurement of the z-spin of an electron which yields the result 'down' projects the spin part of the state-vector of the electron into the state-vector $|-z\rangle$. This sort of ideal measurement is called *maximal*: there is no more spin information about the electron to be had once we know that it is in a certain spin eigenstate.

However, suppose we consider a *nonmaximal* measurement. Suppose we have an ensemble of systems each of which is capable of being in one of at most *three* eigenstates of some observable, say the z-spin of a spin 1 particle. Suppose that the ensemble is a mixture of three pure subensembles, so that its density operator is

$$\rho = \lambda_1|1\rangle\langle 1|+\lambda_2|2\rangle\langle 2|+\lambda_3|3\rangle\langle 3|$$

where $\Sigma_i\lambda_i = 1$.

A device which separates the systems with z-spin $+1$ or 0 ('yes') from those whose spins are -1 ('no') is a measuring device but its result is frequently *nonmaximal*. It does not separate the spin $+1$ particles from the spin 0 particles. The projection postulate tells us that the measurement will separate out the 'yes' subensemble from the 'no' subensemble but apart from this it does not tell us what the conditions of the two new subensembles are. On this point different versions of the projection postulate differ. We examine two projection postulates, the original one due to von Neumann, and one due to Luders. We shall find that Luders' rule gives a better account of ideal minimally disturbing measurements.

The Von Neumann projection postulate

Von Neumann's rule, the rule contained in his discussion of measurement in his treatise,[1] prescribes that a measurement employing the projection operator P_s takes the density operator ρ into the new density operator ρ' where

$$\rho \rightarrow \rho' = P_s/\text{trace}(P_s).$$

Talk about 'the collapse of the wave-packet' suggests either a physical process undergone by an individual quantum system which 'collapses' or a change in the most appropriate description of a pure ensemble, namely a new pure ensemble. The projection postulate is therefore a generalization of the idea of the collapse of the wave-packet, a generalization which covers both wave-functions, state-vectors and pure ensembles on the one hand and mixtures on the other.

To illustrate von Neumann's rule we consider some specific cases.

In our nonmaximal measurement, filtering out the subensemble of particles having spin $+1$ or 0 from those having spin -1, von Neumann's projection postulate tells us that after the measurement the first subensemble becomes *randomized* between z-spin $+1$ and z-spin 0 particles:

$$P_s = |1\rangle\langle 1| + |2\rangle\langle 2|.$$

Trace $[P_s] = 2$ so that according to von Neumann's rule

$$\rho' = \tfrac{1}{2}|1\rangle\langle 1| + \tfrac{1}{2}|2\rangle\langle 2|.$$

Think of von Neumann's rule as *the clumsy experimenter's rule*. The clumsy experimenter manages to randomize the ensembles that result from nonmaximal measurements for mixtures. The careful experimenter would preserve the weightings of the components in the new subensembles.

In what way does von Neumann's rule collapse a wave-packet? If the initial ensemble is pure, described by the superposition

$$|\,\rangle = c_1|1\rangle + c_2|2\rangle + c_3|3\rangle$$

then a system in the 'yes' subensemble will be exactly the same *mixture* as before, namely

$$\rho' = \tfrac{1}{2}|1\rangle\langle 1| + \tfrac{1}{2}|2\rangle\langle 2|.$$

This is even worse. Von Neumann's rule treats all mixtures and all pure states in the same way, precisely because the final density operator ρ' is independent of the initial density operator ρ.

This is pretty unsatisfactory behaviour if we want to describe ideal measurements. We want a state-vector like a pure ensemble described by the density operator

$$\rho = |\,\rangle\langle\,|$$

above to come out, after filtering, as a mixture of two pure subensembles. The pure subensemble which passes the filter with 'yes' should have the new density operator

$$\rho' = |\,\rangle_{\text{new}}\langle\,|_{\text{new}}$$

with coefficients for $|1\rangle$ and $|2\rangle$ in the new state-vector $|\,\rangle_{\text{new}}$ in proportion to c_1 and c_2. In other words, we want a minimally disturbed state-vector as a result.

Von Neumann's version of the projection postulate, even if it fails with nonmaximal measurements, handles maximal measurements as one would expect, projecting state-vectors on to the state-vector corresponding to the projection operator, turning pure ensembles into pure (sub)ensembles and extracting pure subensembles from mixtures.

If von Neumann's rule is the randomizing, 'clumsy experimenter's rule', the alternative, Luders' rule,[2] is the rule of the ultracareful experimenter.

Luders' rule

Luders' rule gives a nonrandomizing transformation of the old density operator due to measurement. It is in this sense better, at least as a description of ideal measurements, than von Neumann's rule though it is also more complicated. According to Luders' rule the old density operator ρ goes into ρ', where

$$\rho \to \rho' = P_s\,\rho\,P_s/\text{trace}\,(\rho P_s).$$

For the nonmaximal filtering considered above Luders' rule tells us that the density operator for the mixture of z-spin $+1$ and z-spin 0 particles should be

$$\rho' = (\lambda_1|1\rangle\langle 1| + \lambda_2|2\rangle\langle 2|)/(\lambda_1+\lambda_2)$$

which represents a mixture which preserves the weighting of the two original pure subensembles of z-spin $+1$ and z-spin 0. One can think of it as a filtered, but otherwise undisturbed, version of the old mixture.

Similarly, if a quantum system is in the superposition

$$|\,\rangle = c_1|1\rangle + c_2|2\rangle + c_3|3\rangle$$

then a system in the 'yes' subensemble will have the state-vector (up to a complex factor of modulus 1)

$$|\,\rangle = c_1/\surd(|c_1|^2 + |c_2|^2)|1\rangle + c_2/\surd(|c_1|^2 + |c_2|^2)|2\rangle.$$

So here we have two versions of the projection postulate. Neither is simply right or wrong. Each is suitable or unsuitable for a given measurement depending on whether that measurement is randomizing or not. On the other hand, there is no doubt that Luders' rule captures better our intuitions about what an 'ideal measurement' should do to an ensemble. Furthermore, Luders' rule has an interesting connection with the conditional in quantum logic as we shall see in Chapter 10, which is why we have gone into it in such detail.

So far all we are doing is describing the puzzling fact that in quantum mechanics there are two kinds of process, measurements and nonmeasurements, a fact not to be confused with there being two versions of the projection postulate. The two versions of the projection postulate are two accounts of the deviation that measurement makes from Schrödinger's equation. We don't know what accounts for the difference between measurements and nonmeasurements.

Two kinds of process

We need a projection postulate if we are going to have the repeatability of ideal measurements together with the probability interpretation of the state-vector. However, we do know how the projection takes place, or even whether quantum mechanics can properly describe the measurement process, given that a projection does take place.

The fundamental issue in the quantum theory of measurement is: can there be one? Another fundmental issue is: should there be one? Bohr, for one, seemed to have no interest in the merely 'formal' question of whether a quantum theory of measurement could be consistent. We sidestep this worry and examine the first quantum theory of measurement which was due to von Neumann.

Von Neumann's is perhaps less a theory of measurement than a set of results about the conditions such a theory must satisfy. However, these results generate puzzles of their own and are the focus of what philosophers of quantum mechanics call the measurement problem. In this limited sense then there certainly can be a quantum theory of measurement.

Von Neumann begins by supposing that we can describe both the measured system S and the measuring apparatus M via the formalism of quantum mechanics. We suppose that we can assign state-vectors to both M and to S, a supposition that Bohr's version of the Copenhagen interpretation would not allow. We ask: what makes the interaction of M and S a measurement?

The minimal answer is that M registers a result which must be *correlated* with whatever state S has *after* the measurement interaction. The result registered by M must reveal this state, otherwise there would be an interaction but no measurement.

In the case where S is initially in an eigenstate of the observable (say **A**) that M measures, the measuring process need not alter the state of S. So let this state of S, both before and after the interaction, be $|S, k\rangle$, and let the eigenstates of M corresponding to each of the possible results a_i of the measurement of **A** be $|M, i\rangle$.

Suppose that the measuring apparatus M is in state $|M, j\rangle$ just before the interaction with S. If the interaction is a measurement then the joint state of $S+M$ is represented by a state-vector in the *tensor product* of the two Hilbert spaces associated with S and M. It evolves as follows.

$$|S, k\rangle \otimes |M, j\rangle \rightarrow |S, k\rangle \otimes |M, k\rangle.$$

In other words, the apparatus evolves from state j to the state k exhibiting the result that S is (and was as it happens) in the state k. That there are Hamiltonians describing couplings between S and M which create this effect, of changing M without disturbing S, follows from von Neumann's theory. The art of measuring the quantity **A** is the art of constructing interactions between S and the chosen M which generate this evolution of M.

The more general case, when S is initially *not* in an eigenstate of the measured observable **A**, is the more interesting. One can naturally represent whatever state S is in as a superposition of the states $|S, k\rangle$, the eigenstates of the observable that M measures. The measured system S will therefore initially be in the state

$$|S\rangle = \Sigma_k c_k |S, k\rangle$$

where $\Sigma_k |c_k|^2 = 1$.

By linearity the previous theory applies, so that the evolution is this time

$$(\Sigma_k c_k |S, k\rangle) \otimes |M, j\rangle \rightarrow \Sigma_k (c_k |S, k\rangle \otimes |M, k\rangle.$$

In other words, the combined systems $S+M$ finish the interaction in a state represented by a state-vector in the tensor product of their individual Hilbert spaces. From the point of view of S and M separately this is a funny state. Neither S nor M can be said to have its own state. Neither S nor M is separately a superposition of states. The measuring apparatus M is certainly not in a determinate state. Yet we make measurements all the time and find that the measuring apparatus is always determinate. How can we reconcile these results, the determinateness of the apparatus and its nondeterminateness according to the Schrödinger evolution of the combined systems?

Von Neumann's answer is that we cannot reconcile them.

He distinguishes between two kinds of evolution a quantum system can undergo. In processes of the first kind the change is discontinuous and in violation of Schrödinger's equation. By contrast processes of the second kind are governed by Schrödinger's equation and the changes of state they induce are continuous. Von Neumann also proves a simple but important 'consistency' result, namely that if a further measuring system interacts with the first measuring system and measures the value of the dynamical variable of S displayed by the first apparatus the result displayed by the second system will be consistent with that registered by the first apparatus. The process can be iterated *ad infinitum*.

Now we face an awkward problem: if there are two kinds of process what makes the difference between them? Almost every answer to this question in the literature strikes most people as bizarre.

One answer is that measurements are made special by the consciousness of an observer.[3] The observer's consciousness is presumably to be thought of as a sort of stuff which resides at the end of a causal chain beginning with the measured system and connecting the measuring apparatus, the observer's nerve endings and brain processes. On this view the world is indeterminate until viewed. Viewing throws the observed system and the apparatus used to measure it into determinate states. We are back with Schrödinger's Cat. All this applies just as much to everyday interactions as to those artificial ones we call laboratory measurements. Therefore the proponent of this view thinks of consciousness as constantly intervening in the world to transform it from a fuzzy world into the sharp world we live in.

But then one is inclined to ask whether all forms of consciousness can do the trick. Can dogs do it? Mice? Insects? Can I do it even when I am day-dreaming and not properly attending to the world I am measuring?

Furthermore, are we really to respond to a difficulty in a rather successful physics theory by proposing a revolutionary new form of Idealism? Something of this objection applies to the next suggestion.

Another incredible answer, that of the *many-worlds interpretation*[4] which is taken very seriously by a few physicists and philosophers and not at all by most others, is to deny von Neumann's conclusion that there are two types of processes and to claim that upon each measurement the universe splits up into myriad new universes. In a sense, the many-worlds interpretation preserves determinism, continuity and Schrödinger's equation and avoids the need for a 'cut' between observed and observing systems. It also allows one to speak of the 'state-vector of the universe' and so is of interest to cosmologists and to students of quantum gravity, synthesizers of general relativity with quantum theory.

Against the many-worlds interpretation one can say that it is far from clear.[5] For example, it is not clear whether at each measurement the whole universe is to be thought of as splitting into many equally real noncommunicating universes. If it is we have an extreme violation of the principle of the conservation of mass–energy. Of course we still have one universe at least in the sense that all the new universes can be comprehended under one heading. Therefore it is not clear that the 'many-worlds' label is accurate.

In this realistic version of the interpretation we imagine the universe as cascading into new universes. Every time a measurement is made new copies of you and your body appear in myriad new worlds. Why does the particular you that you think of having your biography appear in just *this* branch of the cascading universes?

If on the other hand we think of the states as splitting into new *possible worlds* each said to be real but only one of which is said to be 'actual', that is if we think of these states as 'real' in the way that the metaphysically modal realist thinks of possible worlds as real, then apart from the metaphysical gloss the difference between this interpretation and the ordinary interpretation seems to be negligible.

The many-universes interpretation offers a choice between a truly bizarre reading of quantum mechanics which retains a kind of chanciness (why am *I* in *this* world?) with the appearance of determinism and a modal reworking of talk of collapsing wave-packets.

There is a way of denying that there are processes of the first kind and retaining determinism without appealing to many universes. In general the measurement of an observable will transform a pure state into a mixture. This happens when the pure state is not an eigenstate of the measured observable. But suppose that his transformation of a pure into a mixture does not really happen but only appears to happen. Suppose that it could be shown that a continuously evolving pure state and the mixture are *statistically equivalent,* giving the same expectation values for all *macroscopic* observables. Then one could argue that there is no reason, other than convenience, for appealing to the collapse of the wave-packet in measurement.

Daneri, Loinger, and Prosperi,[6] in their thermodynamic theory of measurement, showed that superposition and mixture are statistically equivalent for macroscopic observables applying to macroscopic systems. The interference terms which characterize the superposition get washed out for macroscopic observables. This seemed to some philosophers, like Nancy Cartwright at one time, to offer a way out for the realist who wanted the universal truth of Schrödinger's equation. Mixtures might be used in describing the results of measurements but only as a matter of convenience. Of course all this can be of comfort only to an ensemble interpretation. If you take quantum mechanics to apply to the individual system you can expect the interference terms to show up in the usual way.

However the statistical equivalence of superposition and mixture does not guarantee that the behaviour of individuals in the ensemble will be equivalent. Schrödinger's Cat provides the obvious example. Suppose that after one hour half the cats are dead, so that after two hours three-quarters are. The superposition principle allows some of the cats which are dead after one hour to be alive after two hours. Only the proportions of dead and alive cats are the same in mixture and superposition. But macroscopic systems have orderly histories,[7] as Cartwright now says, and so, whatever the merits of the Daneri–Loinger–Prosperi theory of measurement, the superposition description has to go.

One theory of the measurement process explains how the chance collapse of the wave-packet occurs (it's an effect of the mind), another explains how it appears to occur while what really happens isn't chancy, another explains how it appears to occur when it doesn't occur at all. A fourth claims that the chancy collapse happens much more often than we think. This is Nancy Cartwright's view, presented in her book *How the Laws of Physics Lie*.[8]

Why think that the collapse of the wave-packet happens all over the place, and not just as a result of measurement? Cartwright cites the case of the Stanford Linear Accelerator Center (SLAC). In SLAC electrons surf down the accelerator on an electromagnetic wave, approaching the speed of light after a few miles. But as soon as a free (or a surfing) one is localized its wave-packet will spread out through the whole universe. To sustain the surfing picture the electron has to be continuously localized or collapsed.

The trouble with Cartwright's view is that, like particle interpretations of quantum mechanics, there is a whole range of phenomena which it cannot explain. Thus Cartwright has nothing to say about EPR, which demands that a collapse does *not* occur until a measurement is made. So here again is the usual trade-off between a simple and sensible explanation of some quantum effects, and the complete failure to explain others.

Should there be a formal quantum theory of measurement at all? Bohr's view seems to have been that there should not. At least he evinced almost no interest in the subject beyond what could be conveyed by epistemological analysis. This is, ironically, one strength of Bohr's antirealism, that it can avoid some puzzles altogether. Which is better, a bold hypothesis which encounters paradox or a weak one which doesn't?

The ignorance interpretation of mixtures

Connected with the process of making measurements on mixtures is another puzzle: How seriously do we take the descriptions we give of mixed ensembles? Or, in other words, how should we interpret the density operator and the neat and elegant density operator formalism?

A very modest extension to the minimalist statistical interpretation, though not one acceptable to Ballentine or Popper, consists in claiming that when an ensemble is pure, all the elements of the ensemble are in the same state. In other words, quantum mechanics, which describes ensembles, does not describe the individual system, *except* in the case when the ensemble is pure. In that case the state of the ensemble can be thought to devolve on to its elements.

How is this idea relevant to *proper* mixtures?

Suppose you mix two beams, the first a pure beam of spin 'up' in the x-direction, and the second a pure beam of spin 'down' in the x-direction. Then you will want to say that the new mixture consists of two subbeams, one with all its members having spin 'up' in the x-direction, and the other with all its members having spin 'down' in the x-direction. The mixing of the two subensembles does not interfere with the two subensembles except to make them parts of a larger whole.

The density operator for the ensemble is

$$\rho = \tfrac{1}{2}|+x\rangle\langle +x| + \tfrac{1}{2}|-x\rangle\langle -x|.$$

It looks as though you can read off from ρ the fact that half the electrons have spin 'up' and half have spin 'down' in the x-direction and none has spin 'up' in any other direction. This follows from the *ignorance interpretation of mixtures,* according to which

(1) whenever we know that a quantum system belongs to a mixture, it really belongs to one of the pure subensembles of which the mixture is composed, though we do know which;

(2) the coefficients in the expression for the density operator give us a measure of our knowledge of which subensemble it does belong to.

The ignorance interpretation of mixtures takes the mixing of quantum-mechanical ensembles to be just like the mixing of classical ensembles. Probability in quantum mechanics is then made up of two components, one purely quantum mechanical, a result of the superposition principle and affecting even pure ensembles, the other applicable both to quantum and classical mechanics, a matter of a mixing of pure ensembles.

The trouble is that the density operator ρ can be represented as

$$\rho = \tfrac{1}{2}|+d\rangle\langle +d| + \tfrac{1}{2}|-d\rangle\langle -d|$$

for any other direction d.

Therefore, according to the ignorance interpretation of mixtures, we could also infer that half the electrons in the beam have spin 'up' and half spin 'down' in the different d-direction. However no electron can have determinate spins in two directions, the x-direction and the d-direction.

One possible way out for the ignorance interpretation is to drop its claim (2), that we can read off the composition of a mixture from its density operator. Thus, not all representations of the density operator may be equally acceptable, even though all are equivalent statistically. There may be a preferred description, perhaps one that recognizes how the mixture was put together. However this has the unfortunate consequence that the quantum-mechanical formalism is subject to a new type of incompleteness.

This puzzle about how to take the density operator is important in quantum mechanics because it is about the quantum-mechanical concept of state, and it is a measure of the peculiarity of quantum mechanics that this is a slippery notion. There is a way to rescue an ignorance interpretation of mixtures, at least an ignorance interpretation of sorts. We make use of quantum logic to do just that in Chapter 10.

8

Nonlocality and hidden variables

Like the superposition principle, the indeterminacy relations and the complementarity theorem, nonlocality, broadly construed, is at the heart of even the very simplest applications of quantum mechanics. Locality means, among other things, 'no action-at-a-distance'. It means that the properties of a physical system are affected only by events in the immediate vicinity of the system. It also means that complex physical systems may be described as collections of interacting, but otherwise independent, components. *Non*locality can be a matter of nonlocal forces, of nonlocal correlations, or of physical holism.

We speak ambiguously of nonlocal effects, a nonlocal universe and nonlocal theories. Newton's theory of gravitation is a nonlocal theory. There, nonlocality is a matter of instantaneously transmitted forces. In the second and third of these senses of locality (insofar as they are really separable) – nonlocality without forces – quantum mechanics is a nonlocal theory.

If you naively identify an individual quantum system with the square modulus of its wave-function, as Schrödinger did, you will find that a successful position measurement localizing the system to *here* forces its total disappearance from *there* at the instant of measurement. Immediately after the measurement the system will reappear *everywhere* since the position operators r_t and $r_{t'}$ are totally incompatible when $t \neq t'$. This result follows from the complementarity theorem. Instantaneous disappearing and reappearing would involve an action-at-a-distance and would violate special relativity, as Einstein pointed out in 1927. However, perhaps this sort of nonlocality is simply too naive, relying as it does on the projection postulate and on associating the system too literally with its wave-function.

On the other hand imagine, as Born did originally, that the individual system is a nicely localized particle. Then again you run into nonlocality. The obvious example is provided by the two-slit experiment. How can the slit the particle does not go through affect where it hits the screen behind the diaphragm unless there is action-at-a-distance?

The Pauli exclusion principle provides a naive example of holistic nonlocality. If electrons are localized particles there must be some nonlocal mechanism which stops the two electrons in (for example) a helium atom occupying the same state. On a particle view of electrons, chemistry is possible only because of the nonlocal structuring of electrons in complex atoms.

Philosophers seem to have been strangely unimpressed by the peculiar nature of the exclusion principle, though here Margenau[1] is an honorable exception. The exclusion principle tells us that no two fermions – spin-$\frac{1}{2}$ particles like protons, neutrons and electrons – can be in the same state.

However, take two electrons in *two adjacent* hydrogen atoms. Can't they both be in the ground state with spin 'up'? Of course, they can. So that under one description – as two electrons in two different atoms – they are in the same state. This is ruled out by the exclusion principle, so we should really say that the two adjacent atoms form a single system, and that within this system the two electrons are in different states. Quantum-mechanical talk of the *two* electrons really requires that we think of both electrons as part of one system which has, along with the two electrons, two protons spaced apart by the distance that separates the hydrogen atoms. Now the two electrons are in two different states in this single overall system. Quantum mechanics requires us to take a *holistic* view of the two-hydrogen-atom-whole. Imagining, as we do, single hydrogen atoms, and solving the Schrödinger equation for the electrons in them, is exactly the sort of 'abstraction' that Bohr warned against in his Como lecture.

As a 'formal, regulative device of great generality',[2] the exclusion principle requires us to assign an antisymmetric state-vector to a group of identical fermions. The state-vector must change sign under exchange of a pair of identical fermions. Reflecting on the two-hydrogen-atom case we might be disposed to the view that quantum mechanics implies that two electrons *lose their identity* in a two-hydrogen-atom-whole.

If this were so we could not even formulate the requirement of antisymmetry – change of sign under particle exchange. To put it another way, because we can and must be able to *refer* to the two electrons, if we are to formulate the antisymmetry requirement, they still retain their separate identities. This is puzzling, since if they still are distinct, we are tempted to think that they avoid one another's states via a nonlocal interaction.

If there is a straightforward lesson here it is that naive pictures of quantum systems – as being either waves or particles – lead straight to the nonlocality of quantum mechanics. The genius of EPR is that it demonstrates *without naivety* that either quantum mechanics is nonlocal or that it is incomplete. Einstein, we recall, assumed that nonlocality was a necessary feature of any sensible physical theory and so argued for the incompleteness of quantum mechanics. If quantum mechanics is incomplete then it should presumably be supplemented with a deeper 'hidden-variables' description of microphysical states and one would expect that quantum-mechanical states would turn out to be mere averages over these deeper states.

The most important contribution to the philosophy of quantum mechanics since EPR – J. S. Bell's work on his inequalities – consists of a deepening and generalizing of the EPR, or rather the EPR–B thought-experiment. If EPR showed that quantum mechanics is either nonlocal or incomplete, Bell shows that it is nonlocal anyway.

EPR–B considers correlations between spin measurements in the same direction. For the singlet state of a pair of spin-$\frac{1}{2}$ particles these spins will be perfectly anticorrelated, that is, always oppositely directed. Bell asks the more general question: how will the results of spin measurements on such a system be correlated when you make two measurements not in the same direction, but in *any pair of directions, at any angle with respect to one another* that you like? Furthermore, how *must* they be correlated if the results are governed by two sets of hidden variables, each set applying separately to each member of the pair? How, in other words, must they be correlated if a deterministic, *local* hidden-variables theory truly describes them?

The Bell inequalities

Bell's answer is that the spins of particles controlled by a deterministic, local hidden-variables theory, must be correlated in such a way as to satisfy certain mathematical relations called the *Bell inequalities*.

There is in fact a family of Bell inequalities. Bell inequalities, though they are the result of thought-experiments, can be tested by real experiments and much work in the 1970s was directed to finding out if quantum mechanics violated them. Some of them are more easily testable by experiment than others. Some may be more easily mathematically derived than others. We begin with a very elegant derivation due to P. H. Eberhard[3] of one of the Bell inequalities, which is as follows.

Imagine, as in the EPR–B thought-experiment, a source of correlated pairs of electrons I and II, each pair being in the singlet spin state, the source being placed between and separated from the detectors by a large distance. Each of the two detectors can measure the spin component of one of the two electrons in any chosen direction. The only difference between the present example and that of EPR-B is that the spin measurements on I and II need not be made in the same direction.

In fact we have to consider *two* directions for the measurements on I, let these be a and a', and *two* directions for the measurement on II, let these b and b'. Therefore one can measure any one of *four* possible combinations of spin measurements on any given pair: a or a' for I and b or b' for II.

More interestingly, by making a large number of measurements for each combination of directions a, a', b, b' one can measure the *correlation* between the spin measurements on I in a given direction and on II in another direction. There are only two possible results of a spin measurement, 'up' and 'down'. Let the direction of the spin measurement on I be a and on II be b. Electron I may be found to have spin 'up' ($+1$) or 'down' (-1) in the direction a and electron II spin 'up' ($+1$) or 'down' (-1) in the direction b. These results may be positively correlated, negatively correlated or not correlated at all, depending on the directions a and b. So we need to define a measure of their correlation, or lack of it.

The first question is therefore: how are we to define the correlation between the results of the measurements on I in one direction and on II in (possibly) another direction?

Consider some special cases. Suppose that the measurements on I and II always give the same result, that is if I is 'up' then so is II, and if I is 'down' then again so is II. Intuitively the correlation is perfect, that is $+1$. One can determine the correlation $C(a,b)$ as follows. Make a large number N of the spin measurements on I and II. For each pair multiply the two values of spin. Form the sum of the products for the N pairs, and divide by N. In other words, if a_n is the result of the nth measurement on I in the direction 'a', and b_n is the corresponding nth measurement on the system II correlated with I in the direction b, then, as N goes to infinity

$$C(a,b) = (1/N)\, \Sigma_n a_n b_n.$$

It is easy to see that if the spins are perfectly correlated in the directions a,b, then $C(a,b) = +1$. If they are perfected anticorrelated then $C(a,b) = -1$. If they are wholly uncorrelated, then $C(a,b) = 0$. If the two measurements are made in the *same* direction, then quantum mechanics tells us that the spins of I and II are always oppositely directed. So for correlated electrons $C(c,c) = -1$, for any direction c chosen for the measurements made on both I and II.

The mathematical problem is this: find an inequality relating $C(a,b)$, $C(a,b')$, $C(a',b)$, and $C(a',b')$ which is satisfied by any local hidden-variables theory but not by quantum mechanics. We proceed as follows.

The first step is to consider measurements on the nth correlated pair, I and II.

Let the result that *would be obtained* if the spin of I *were measured* in direction a be a_n. This will be either $+1$ or -1. Similarly, let the result that would be obtained if the spin of I were measured in direction a' be a'_n. This again will be either $+1$ or -1.

Of course one cannot measure both these spins without disturbing the system. But one supposes that the result of the unmeasured spin is determinately either $+1$ or -1. This is a perfectly correct assumption given that I and II are controlled separately by a deterministic, hidden-variables theory.

One now defines similar quantities b_n and b'_n for the correlated system II.

Now consider the odd-looking quantity

$$g_n = a_n b_n + a_n b'_n + a'_n b_n - a'_n b'_{n'}. \qquad [*]$$

We manipulate g_n to obtain the desired inequality. But first note that one can rewrite g_n as

$$g_n = a_n(b_n + b'_n) + a'_{n'}(b_n - b'_n)$$

The key points are these.

Since b_n and b'_n are each either $+1$ or -1, then either they have the same sign or different signs. So either

$$(b_n + b'_n)$$

or

$$(b_n - b'_n)$$

is zero. The first is zero if b_n and b'_n are different in sign, the second is zero if b_n and b'_n are the same in sign. The one which is not zero is either $+2$ or -2. Since a_n is either $+1$ or -1 it follows that $g_n = +2$ or -2. Since a_n is either $+1$ or -1 it follows that $g_n = +2$ or -2. Hence, and lastly, the *absolute value* of g_n, that is $|g_n|$, is 2.

We want to get an expression involving $C(,)$ rather than g_n. So from [*] we can write

$$(1/N)\Sigma_n g_n = (1/N)\ \Sigma_n\ (a_n b_n + a_n b'_n + a'_n b_n - a'_n b'_n)$$

and taking absolute value of both sides, we get

$$|C(a,b)+C(a,b')+C(a',b)-C(a',b')|\leqslant 2 \qquad \text{(BELL-1)}$$

This is one of the Bell inequalities.

Quantum mechanics violates it. It turns out that the world violates it. This does not show that quantum mechanics is true. It does show that the classical world picture is irretrievable.

The quantum-mechanical correlation $C(a,b)$ for spin measurements made on two correlated electrons in the singlet pair state is easily shown to be $-\cos\,\theta_{ab}$ where θ_{ab} is the angle between a and b. This accords with our intuition, as far as that goes. If a and b are in the same direction the correlation is -1, if a and b are oppositely directed the correlation is $+1$ and if a and b are a right-angle apart the correlation is zero.

One can get this quantum-mechanical result as follows. Let a be the z-direction, and let the angle between a and b be θ_{ab}. Make measurements of spin on I in the direction a. If, for the nth particle the result is 'up' (+), probability for the 'up' result to a measurement of spin on II in direction b is

$$\sin^2 \tfrac{1}{2}\theta_{ab}.$$

If the result of the measurement on particle I is 'down' ($-$), then the probability of and 'up' result for particle II in the original direction b is

$$\sin^2 \tfrac{1}{2}\pi-\theta_{ab}.$$

This second result represents an *anti*correlation. So taking away the second from the first and dividing by two, since these are two equally likely outcomes, we get

$$\tfrac{1}{2}\,(\sin^2 \tfrac{1}{2}\theta_{ab}-\sin^2 \tfrac{1}{2}\pi-\theta_{ab}) \;=\; -\cos\,\theta_{ab}.$$

To violate the inequality (BELL-1) choose a and b to be in the same direction, say vertically upwards. Let a, a', b and b' all be in the same plane and let a' be 45° to the left of a and b' be 45° to the right of b, so that the angle between a' and b' is 90°. Substituting this case into inequality (BELL-1) and using the facts that

$$\cos 45° \;=\; 1/\sqrt{2}, \quad \cos 90° \;=\; 0$$

we have

$$|-1-1/\sqrt{2}-1/\sqrt{2}-0|\leqslant 2$$

or

$$1+\sqrt{2}\leqslant 2$$

which is an arithmetic contradiction.[4]

First, one should not imagine that nonlocal quantum correlation is an oddity that afflicts only trumped up thought-experiments. The correlated electron pair may constitute a toy example, but it has been the subject of discussion only because it is so simple to handle mathematically. Quantum correlations will arise whenever quantum systems interact according to conservation laws, which is all the time.

Second, the Bell inequalities are testable. The experiments performed to date use the formally similar set-up of polarized photon correlation generated during the decay of an atom from an excited state. The experiments are difficult to perform because photon detectors and polarizers are not perfectly efficient, and their inefficiency makes the small statistical deviations due to quantum nonlocality difficult to detect. But the current consensus is that the Bell inequalities have been shown to be false *experimentally* for some quantum correlations. Therefore, even if quantum theory turns out to be false, its successor will do nothing for the classical world picture.

Finally, it does not follow that one can send a *message* from a person **A** located at the apparatus used to make a measurement on I to a person **B** located at the apparatus used to make the measurement on II. On the other hand, the behaviour of the two people can be correlated by the measurements on an electron pair.

Persons **A** and **B** can agree to act according as each gets spin 'up' or 'down' for a particular pair of measurements. If these measurements are in the same direction, then the actions of **A** and **B** will be perfectly correlated. But they would be equally correlated if they were measuring correlated classical systems, say correlated spinning tops. If **A** and **B** act conditionally of the results of spin measurements made in different directions, their correlations will be different from any they could make using classically spinning tops, but the fact that they are correlated at all is no more paradoxical than in the classical case.

No-hidden-variables theorems

A demonstration that quantum mechanics is inconsistent with an underlying hidden-variable theory predates Bell's work by more than thirty years. In 1932 von Neumann showed that no state of a quantum system can assign simultaneously definite values to all the quantum-mechanical observables. The arguments put forward by von Neumann employed a set of assumptions at least one of which was highly questionable. Nevertheless, his prestige, the obscurity of the empirical significance of his proof, and the fact that there was a handy philosophy which denied the possibility of hidden-variables theories all conspired to generate an anti-hidden-variables orthodoxy which survived Bohm's hidden-variables theory of 1952.

In his proof von Neumann assumed that every Hermitian operator on a Hilbert space of dimension greater than one represents an observable, that for a particular vacuous observable **1** – the identity – the expectation value

$$\langle \mathbf{1} \rangle = 1,$$

that for each observable **A** and each real number r

$$\langle r\mathbf{A} \rangle = r\langle \mathbf{A} \rangle,$$

that if **A** is an observable whose value is always greater than or equal to zero then the expectation value of **A** should also be greater than zero.

These assumptions are innocent enough but von Neumann assumed a fifth, which he assumed to hold also in strong form for the hidden variables, namely that for arbitrary observables **A, B, C** ⋯

$$\langle \mathbf{A}+\mathbf{B}+\mathbf{C}+\cdots \rangle = \langle \mathbf{A} \rangle + \langle \mathbf{B} \rangle + \langle \mathbf{C} \rangle + \cdots$$

This enables him to prove that there is always an observable **X** such that

$$\langle \mathbf{X}^2 \rangle \neq [\langle \mathbf{X} \rangle]^2.$$

In other words for any quantum state there is an observable which is not dispersion-free, an observable which, in other words, is not definite in value.

It is true that in quantum mechanics we do accept the fifth assumption in its weaker form as applying to quantum-mechanical operators. We do this when we calculate the expectation of position and momentum for the electron in a hydrogen atom. However, one can argue that the assumption is objectionable in that it might not apply to the *hidden-variable states* which underpin quantum mechanics. If it does hold in the following form, namely that if **h** is the hidden variable (or variables, in which case think of **h** as a vector), then even for incompatible **A** and **B**

$$(\mathbf{A}+\mathbf{B})\,(\mathbf{h}) = \mathbf{A}(\mathbf{h}) + \mathbf{B}(\mathbf{h}) \qquad [**]$$

an equation which says that the value of **A** + **B** will be determined by the values of the hidden-variables which determine **A** and **B** separately, then it is very easy to derive a no hidden-variables theorem.

But why doubt [**]? Notice that to measure **A** + **B** one cannot simply measure **A** and then **B** and add the results since the two measurements interfere with one another. Thus we know that an electron in the ground state of a hydrogen atom has a definite energy. Its Hamiltonian is the sum of its kinetic and potential energies. But neither of these has a determinate value. If one were to measure its kinetic energy and then its potential energy, the latter from a measurement of its position, and then add up the answers you would not get the ground state energy.

Now for the proof, which is due to Rudolf Peierls.[5]

Consider the spin component s', in a direction θ to the x-axis, of a spin-$\frac{1}{2}$ particle

$$s' = s_x \cos\theta + s_y \sin\theta. \qquad [***]$$

According to quantum mechanics s', s_x, and s_y each take only the values $+\frac{1}{2}$ or $-\frac{1}{2}$. For a hidden-variables theory to reproduce this result we must find functions $S_x(\mathbf{h})$ and $S_y(\mathbf{h})$ which always take the values $+\frac{1}{2}$ or $-\frac{1}{2}$ such that the value of $S'(\mathbf{h})$ given by [***] is also $+\frac{1}{2}$ or $-\frac{1}{2}$, which is impossible (except when θ is a multiple of $\frac{1}{4}\pi$).

To show that von Neumann's result depends on this strong assumption about the hidden variables, Bell produced a hidden-variables theory for electron spin which got the expectation values right.[6]

For this reason the more subtle theorem of Gleason is the more powerful, and has received much more discussion by philosophers of quantum mechanics. The equally well-known Kochen and Specker no-hidden-variables theorem is, incidentally, essentially a special case of Gleason's theorem. Gleason proved von Neumann's result, as a corollary to his main theorem,[7] by assuming that the fifth assumption held for pairwise compatible observables **A, B, C,** . . . He therefore strengthened von Neumann's result, at least in this respect. In one other, unimportant respect he weakened it. He derived it only for Hilbert spaces of dimension three or more.

However, even Gleason's result cannot rule out a class of hidden variables according to which the result of making a measurement on a system depends not only on the values of the hidden variables but also on the experimental apparatus used to make the measurement. Such hidden-variables theories are called *contextual* hidden-variables theories. To see this consider three observables **A, B, C.** Let **A** and **B,** and **A** and **C** be compatible. It does not follow that **B** and **C** are compatible. You may be able to measure **A** and **B** using on experimental arrangement, and you may be able to measure **A** and **C** using another. But measuring **A** and **B** simultaneously will disturb the value of **C** even though measuring **A** on its own need not. Here is a case in which expectation values are context dependent. We should consider not simple expectation values like $\langle X \rangle$, but relativized expectation values like

$$\langle X, C \rangle$$

where C is a 'measurement context'.

It may seem that contextual theories are somewhat far fetched, although Bohm's hidden-variables theory of 1952 was contextual, and it may also seem that they depart from the spirit of the hidden-variables idea. However, their possibility shows just how difficult it is to rule out hidden-variables theories completely simply on the basis of the structure of the quantum-mechanical state-space.

For this reason, and because it is so simple, and because it leads to concrete experimental work, Bell's work on nonlocality has come to be considered the most significant contribution to our understanding of the obstacles facing a recovery of the classical ontology via hidden variables.

9

A user-friendly quantum logic

Why should anyone think that what is so deeply puzzling about the quantum world is really a matter of logic? Why should anyone think that studying the *logic of quantum mechanics* should be a route to a proper understanding of quantum mechanics?

There is a tradition in the philosophy of quantum mechanics, a tradition stronger among philosophical commentators than among physicists, which sees the subject as the business of understanding quantum-mechanical language and the way that language relates to the world. It is an idea more strongly held by philosophers – it comes much more naturally to them – because the analytic tradition, in which most Anglo-American philosophers of science are raised, is generally much exercised about words and how they hook on to things.

The analytic tradition in the philosophy of quantum mechanics is not unreasonable. Everyone admits that we cannot picture the microphysical world, that a graphic or iconic representation of quantum systems is impossible. Therefore understanding quantum mechanics must be a matter of understanding the logic of the words and the mathematics of quantum mechanics. If this seems implausible, it is because of an ambiguity in our use of the verb 'to understand'.

We understand the physical world, and we understand physics. But physics is not the physical world. It is something of an entirely different sort, a human product, a way of representing the world. In fact, in so far as the expression 'understanding the world' has any sensible application in the philosophy of physics, it must mean understanding the way physics represents the world. Quantum physics is, ultimately, a network of propositions, not of pictures, diagrams, iconic models or mental images. One qualifies these remarks about understanding with 'in the philosophy of physics', because there may well be a way of understanding Other People (say) which is not a matter of understanding representations or propositions. There may also be a corresponding but irrelevant sense of 'understanding the world', that is 'being streetwise'.

If this view of the philosophy of physics is right, and if understanding quantum mechanics is a matter of understanding quantum-mechanical language, then here is the place to make that characteristic move of contemporary analytic philosophy, what Carnap called the move from the material mode to the formal mode, what Quine calls *semantic ascent,* the 'shift from talk of objects to talk of words'[1], because in the case of quantum mechanics, semantic ascent may be obviously nontrivializing, even if it seems to be trivializing in some branches of philosophy. (Sometimes the trivializing effect is the result of a refusal to get down to real philosophical business, as in the case of some 'ordinary language' philosophy. Sometimes it is worse, as when the serious business of the philosophy of religion gets turned into a study of 'God-talk'.)

If the logic of quantum mechanics is nonstandard then the source of the paradoxes may just be our imposing classical logic on it. Move from talk about the world – one aspect of real physics – to talk of talk about the world – the philosophy of physics – and examine the logic of quantum-mechanical language and it may be that the paradoxes dissolve. These last two chapters are taken up with this idea, that quantum mechanics baffles us because we misunderstand its logic.

But what is the logic of quantum mechanics? There are in the literature various attempts to resolve the paradoxes using nonclassical logic. They divide into two types.

There are those that try to impose on quantum mechanics a logic that does not arise naturally from the formalism of the theory. Such is Reichenbach's interpretation[2] which employs a 3-valued truth functional logic and which is generally admitted to be a nonstarter.

Far more important than artificial many-valued logics are the *quantum logics* that arise naturally in the Hilbert space formalism. The strongest of these logics – strongest in the sense of most closely approximating to classical logic – is the logic which mirrors the structure of the set of closed subspaces of Hilbert space. This quantum logic was first studied by the mathematicians Garrett Birkhoff and John von Neumann in a classic paper of 1936.[3] By *quantum logic* we mean Birkhoff and von Neumann's quantum logic and the various ways in which it may be formulated as a logic.

A lattice of propositions is of course not quite a logic, whatever a logic is. But a lattice of propositions has a structure which is at least very much like the structure of a logic. So if quantum logic really is a logic we should first make it look like a logic. Then we will be in a position to discuss whether it really is a logic.

Making quantum logic look like logic

In making the move from the lattice which is quantum logic to quantum *logic* proper, we should ask first what the relation between them is.

Begin with logic. In a logic we will have a set of well-formed formulae (wffs). Some pairs of formulae *P* and *Q* (say) will be interderivable. *P* will imply *Q* and vice versa. Inter-derivability should be an equivalence relation on the class of wffs, so partition the set of wffs into equivalence classes of logically equivalent wffs. The base set of the lattice which corresponds to the logic is just the set of these equivalence classes.

In a logic you will have connectives like conjunction '&', disjunction '∨' and negation '−'. These are mirrored in the lattice by corresponding lattice operations, the meet corresponding to '&', the join to '∨' and the orthocomplement operation to '−'. The partial ordering relation on the lattice corresponds to the 'entailment' relation which holds between pairs of wffs.

This covers '&', '∨', and '−'. What about the conditional '→', and the equivalence '↔'? It so happens that in the quantum logical case these can be defined in terms of '&', '∨', and '−' just as in the classical case, though in quantum logic the definitions are not quite the same.

Lattices and logics are similar structures though clearly they are not the same thing. First of all, a lattice has nothing corresponding directly to a conditional or an equivalence, though both may be definable in terms of the lattice operations. Second, the entailment relation which a logic sets out to capture need not be restricted to hold only between pairs of wffs. We usually take entailment to be a relation which holds between a *set* of propositions and a single proposition. Therefore the effort of rewriting the quantum logic lattice as a logic is not entirely trivial.

Quantum logic can in fact be fairly easily transcribed as a logic in the usual logical styles – as an axiomatic system, as a sequent calculus and as a natural deduction system. Of these, versions of the latter two may be of some philosophic interest. However, whether the resulting system really is a logic is a much discussed philosophical question we postpone until the next chapter.

Axiomatic systems of logic – logics in the Hilbert–Ackermann style – are usually based on a finite set of axiom schemata in which the conditional connective is made to figure prominently. They require a minimal number of rules of inference, often with *modus ponens* as the only one. But axiomatic systems suffer from the following disadvantages: they tend to focus attention on provable formulae rather than on provably valid arguments, which are the real objects of study for any logic, and the structure of their proofs – sequences of formulae each of which is either an axiom or derived from previous formulae in the sequence (usually by *modus ponens*) – is quite unlike that of real reasoning. Axiomatic systems make logic look like a mathematical theory rather than like something humans do.

A sequent calculus, a logic in the Gentzen style, will occupy an intermediate position in the logical spectrum which runs from axiomatic systems at one extreme to natural deduction systems at the other. A sequent calculus deals with sequents. A sequent is a *set* of premises and a *set* of conclusions. A sequent calculus will have one axiom, which allows as 'given' any sequent having a formula common to both premises and conclusions, and a collection of rules of inference, one for introducing each connective into the premises and one for introducing it into the conclusions. There are usually special rules, called structural rules. These are justified by general features of entailment, such as the fact that it is transitive, a property of entailment expressed in the special rule called 'cut'.

Natural deduction systems remedy both of the faults of axiomatic systems. They eliminate axioms at the expense of introducing many more rules of inference. Each connective – and there tend to be five – ideally has two rules of inference: one for introducing it into a formula and one for eliminating it. You can import additional assumptions into a proof at any time, provided you note the dependence on them of what you derive. The great advantage of natural deduction systems over axiomatic systems is their naturalness: they reflect much better the workings of real reasoning, or at least they are supposed to.

A natural deduction system views entailment as a relation between a *set* of wffs and a *single* wff much as one does 'in everyday life'. This fact is responsible for one of the awkwardnesses from which natural deduction systems tend to suffer. Take the introduction and elimination rules for '&' and '∨' in E. J. Lemmon's natural deduction system[4] for classical logic. The '&-elimination' and '∨-introduction' rules are intuitively symmetrical:

$$\frac{A \,\&\, B}{A} \qquad\qquad \frac{A}{A \vee B}$$

These say: from A & B infer A, and from A infer $A \vee B$. The *duality* of classical logic leads us to expect this sort of symmetry between '&' and

'$\vee$'. However the symmetry breaks down when we consider '&-introduction' and the '$\vee$-elimination' rules. &-introduction is a nice simple rule but '$\vee$-elimination' is anything but.

$$\frac{A \quad B}{A \,\&\, B} \qquad\qquad \frac{A \vee B \quad \begin{matrix} A \\ \vdots \\ C \end{matrix} \quad \begin{matrix} B \\ \vdots \\ C \end{matrix}}{C}$$

'&-elimination' says: From a proof of A and a proof of B, infer a proof of $A \& B$. But '$\vee$-elimination' says this: Suppose you have a proof of $A \vee B$; if you have proofs of C from A and of C from B, then you can infer a proof of C.

It is worthwhile contrasting sequent calculi with natural deduction systems. In a sequent calculus one views the entailment relation as holding between *sets* of formulae. Sequent calculi preserve the symmetry between '&' and '$\vee$' at least for those systems (like classical logic and quantum logic) which are essentially dual. The price you pay is that entailment becomes unnatural. In a sequent calculus the proved objects are *sequents*, items of the form $\Gamma \rightarrow \Delta$ which usually receive one of the following interpretations:

(1) when all the formulae in Γ are true, at least one of the formulae in Δ is true; or

(2) for finite Γ, Δ, whenever any conjunction of all the formulae in Γ is true, any disjunction of all the formulae in Δ is true.

In a sequent calculus for classical logic (1) and (2) are equivalent but for quantum logic they are not. There are a couple of different sequent calculi for quantum logic,[5] each open to one but not both of interpretations (1) and (2).

Natural deduction systems are intrinsically friendlier than sequent calculi. Our user-friendly natural deduction system for quantum logic – which we call NDQL – is based on and similar to Lemmon's system for classical logic. All the rules so far discussed carry over into quantum logic except for '$\vee$-elimination' which is restricted in quantum logic.

So what is wrong with '$\vee$-elimination' in quantum logic?

There is in fact an aspect of the classical rule for '$\vee$-elimination' which is suppressed by the diagram above. '$\vee$-elimination' says this: Given a proof of $A \vee B$ from a set of premises Γ, and proofs of C from A (with additional premises Γ_2), and from B (with additional premises Γ_3), you can infer C from the union of the sets Γ_1, Γ_2, and Γ_3. Quantum logical '$\vee$-elimination' requires that the sets Γ_2 and Γ_3 be empty. Quantum logic is dual (in a sense we define below) even though it appears that something has been done to '∨' but not to '&'.

Other classically valid rules succumb in quantum logic. It is useful to explain why, and later on we do just that. For example, if you can deduce a contradiction from a set of premises, then the premises cannot all be true. So take one out, and assume the remainder are true. Then it follows that the one you took out must be false. This is one version of the classical rule of *reductio ad absurdum* (RAA). However, quantum logically you are allowed only sets of premises that have exactly *one* member. If this seems odd, there is an explanation for it.

Similarly, we must restrict the classical rule of conditional proof, that if from a set of premises plus a further premise A you can deduce a conclusion C, then you can deduce the conditional conclusion $A \to C$ from the original set of premises. In quantum logic you are only allowed to apply the rule when the set of premises is *empty*.

Sequent calculi may be contrasted with natural deduction systems in some respects, while in other respects they are quite similar to one another. Sequent calculi are the more elegant and are more useful for proving results *about*, as opposed to proving sequents *within* the system, but natural deduction systems are the more natural, and so we proceed first with a natural deduction system for quantum logic. There is some small philosophical interest in sequent calculi, so we examine them briefly later.

NDQL: quantum logic made easy

We represent sentences in ELQM using lower-case letters a, b, c as propositional variables ranging over the basic sentences of ELQM and using the operators $-$, &, $\vee$, $\to$ and $\leftrightarrow$ to represent the corresponding connective. The class WFFS of well-formed formulae of this, the propositional language of quantum logic, can be defined recursively in the usual way by saying that

(1) propositional variables are wffs;
(2) if a is a wff, so are (a) and $-a$; and
(3) if a and b are wffs, so are $(a \leftrightarrow b)$, $(a \to b)$, $(a \vee b)$, and $(a \,\&\, b)$;
(4) there are no other wffs.

We then go on to say that the connectives are ordered in terms of 'binding power' from $-$ down to $\leftrightarrow$, a fact which enables us to eliminate brackets without ambiguity. This approach can lead to tedium as it does in Lemmon's book.[6]

Better therefore to define the grammar properly in the first place, which we do, for the benefit of the computer literate using a BNF grammar. Thus

$$
\begin{aligned}
\langle\text{wff}\rangle &::= \langle\text{conditional}\rangle \,|\, \langle\text{conditional}\rangle \text{ `}\leftrightarrow\text{' } \langle\text{wff}\rangle \\
\langle\text{conditional}\rangle &::= \langle\text{disjunction}\rangle \,|\, \langle\text{disjunction}\rangle \text{ `}\rightarrow\text{' } \langle\text{conditional}\rangle \\
\langle\text{disjunction}\rangle &::= \langle\text{conjunction}\rangle \,|\, \langle\text{conjunction}\rangle \text{ `}\vee\text{' } \langle\text{disjunction}\rangle \\
\langle\text{conjunction}\rangle &::= \langle\text{negation}\rangle \,|\, \langle\text{negation}\rangle \text{ `\&' } \langle\text{conjunction}\rangle \\
\langle\text{negation}\rangle &::= \langle\text{factor}\rangle \,|\, - \langle\text{negation}\rangle \\
\langle\text{factor}\rangle &::= \langle\text{atom}\rangle \,|\, \text{`('} \langle\text{wff}\rangle \text{ `)'} \\
\langle\text{atom}\rangle &::= \text{`}A\text{'} \,|\, \text{`}B\text{'} \,|\, \ldots \,|\, \text{`}Z\text{'}
\end{aligned}
$$

This BNF grammar automatically gives the usual operator precedence for the connectives. It is equivalent to the more usual grammar, and it allows one to have as many brackets as desired.

We take the class WFFS of wffs which our BNF grammar yields. We need three more basic ideas – those of *sequent, rules of inference,* and *proof.*

A *proof* in our natural deduction system NDQL is a finite, nonempty sequence of *sequents*. Each sequent in the proof is a pair, the first element of which is a finite (possibly empty) set of wffs, the premises of the sequent, while the second, the conclusion, is a single wff. One represents a typical sequent by

$$\Gamma \vdash a$$

where Γ is a set of wffs and a is a wff. One reads $\vdash$ as 'proves', so the sequent is read 'Γ proves a'. Natural deduction sequents are therefore different from sequents in sequent calculi. The things that are provable in NDQL are natural deduction sequents, as in Lemmon's system for classical logic. When one has proved a sequent of the form

$$\Gamma \vdash a$$

in NDQL one has shown that an argument from the set of premises Γ to the conclusion a is quantum logically valid. If Γ is empty then one has proved a on the basis of no premises at all and so a is a quantum logically provable formula.

A proof in NDQL is a sequence of sequents constrained by *rules of inference*. Abstractly, a rule of inference is a set of pairs. The first element in each pair is a finite (possibly empty) set of sequents and the second is a sequent. This, at a pretty abstract level, is what a rule is. The intention is that each pair in a rule will conform to a pattern. Take '&-elimination' as an example. This consists of a set of pairs like

$$\langle\{\Gamma \vdash A \mathbin{\&} B\}, \Gamma \vdash A\rangle$$

or like

$$\langle\{\Gamma \vdash A \mathbin{\&} B\}, \Gamma \vdash B\rangle.$$

The pattern here is that the second element of each pair is one of the conjuncts of one of the members of the first element in the pair. One says '$\Gamma \vdash A$ (or $\Gamma \vdash B$) follows from $\Gamma \vdash A \mathbin{\&} B$ by "&-elimination" '.

Similarly the rule '&-introduction' consists of a set of pairs each of which is

$$\langle\{\Gamma_1 \vdash A, \Gamma_2 \vdash B\}, \Gamma_1, \Gamma_2 \vdash A \mathbin{\&} B\rangle.$$

So what are the rules?

In NDQL there are ten rules and two definitions.

To get a proof started we need a special rule which enables us to write down any sequent which has a certain special form. The *rule of assumptions* enables us to write down any sequent of the form

$$\{a\} \vdash a$$

at any stage of the proof. In fact every proof in NDQL has to begin with an application of the rule of assumptions.

$$\{a\} \vdash a$$

really tells you nothing, except that from a you can derive a. We take the name rule of assumptions for this from Lemmon[7] but really it should be called the rule of sequents. What one introduces to a proof is not a formula (or premise or assumption) but a sequent.

The rule of assumptions is the most trivial of the rules. Of the remaining nine, six are exactly as in Lemmon's system for classical logic. Only the three rules which 'discharge' assumptions are different in NDQL. In fact the three discharging rules – conditional proof, *reductio ad absurdum* and $\vee$-elimination – must all be restricted in a simple, straightforward way. In NDQL you therefore lose something in comparison with classical logic, but you also gain something – an extra definition for $\rightarrow$ in addition to that for $\leftrightarrow$.

The unchanged rules are the rule of assumptions, *modus ponens, modus tollens,* double negation, &-introduction, &-elimination, and $\vee$-introduction. From now on we drop the curly brackets in the premises of a sequent and for $\Gamma_1 \cup \Gamma_2$ we write

$$\Gamma_1, \Gamma_2$$

and for $\Gamma \cup \{a\}$ we write

$$\Gamma, a.$$

A rule of assumptions

You can always infer $a \vdash a$.

Here is a proof of $P \vdash P$

1 (1) P A

MP modus ponens

From $\Gamma_1 \vdash a$ and $\Gamma_2 \vdash a \rightarrow b$ infer $\Gamma_1, \Gamma_2 \vdash b$.

MT modus tollens

From $\Gamma_1 \vdash -b$ and $\Gamma_2 \vdash a \rightarrow b$ infer $\Gamma_1, \Gamma_2 \vdash -a$.

DN double negation

From $\Gamma \vdash a$ infer $\Gamma \vdash --a$ and from $\Gamma \vdash --a$ infer $\Gamma \vdash a$.

&I &-introduction

From $\Gamma_1 \vdash a$ and $\Gamma_2 \vdash b$ infer $\Gamma_1, \Gamma_2 \vdash a \,\&\, b$.

&E &-elimination

From $\Gamma \vdash a \,\&\, b$ infer either $\Gamma \vdash a$ or $\Gamma \vdash b$.

VI ∨-introduction

From either $\Gamma \vdash a$ or $\Gamma \vdash b$ infer $\Gamma \vdash a \vee b$.

Here are the *three restricted rules.*

CP conditional proof

From $a \vdash b$ infer $\vdash a \rightarrow b$.

According to the classical rule one can infer $\Gamma \vdash a \rightarrow b$ from $\Gamma, a \vdash b$. The quantum logical rule requires Γ to be empty. This prevents one proving such classically valid sequents as

$$\vdash P \rightarrow (Q \rightarrow P)$$

and

$$(P \rightarrow Q), (Q \rightarrow R) \vdash (P \rightarrow R).$$

A corresponding restriction goes for ∨E and RAA.

$\vee E$ $\vee$-elimination

From $\Gamma \vdash a \vee b$, $a \vdash c$, and $b \vdash c$ infer $\Gamma \vdash c$.

RAA reductio ad absurdum

From $a \vdash b \,\&\, -b$ infer $\vdash -a$.

Finally there are *two definitions*. These enable you to replace the conclusion of a sequent with its definition.

$$a \leftrightarrow b = \text{df. } (a \rightarrow b) \,\&\, (b \rightarrow a) \qquad \text{Df.}\leftrightarrow$$
$$a \rightarrow b = \text{df. } -a \vee (a \,\&\, b). \qquad \text{Df.}\rightarrow$$

Like Lemmon, we *represent* a proof in NDQL – a sequence of sequents – as a list of triples, the first element of which is a list of names of premises in the sequent, the second being the conclusion of the sequent and third being the justification for writing down the sequent.

The rules we have given are sound and complete for quantum logic.[8] That is whenever we have

$$a \leqslant b$$

in the lattice which is quantum logic, we have a corresponding

$$A \vdash B$$

in NDQL. (We want to distinguish between what is provable in NDQL, and what is valid in the lattice – its semantics – so we write $A \models B$ for the latter, in the usual way.)

Given the rules, what is the semantics?

We noted that NDQL is sound and complete with respect to the quantum logical lattice which it transcribes. But do we have anything like truth tables?

We are certainly not going to have truth-functional truth tables. The truth tables for quantum logic are going to be incomplete, indeterminate in some places. As a matter of fact we have some latitude in assigning truth values to compound propositions. On the one hand, since the & rules in NDQL are classical we might expect that the classical truth table for & applies to quantum logical &. This leads us to the following truth tables

P	Q	$P\ Q$	$P \vee Q$	$-P$
T	T	T	T	T
T	F	F	T	?
F	T	F	T	
F	F	F	?	

Alternatively, one can say that a sentence of ELQM is true if and only if (iff) the appropriate state-vector belongs to the appropriate subspace, false iff it belongs to the orthogonal subspace and neither true nor false otherwise, and one can make negation look rather more like a truth-functional connective at the expense of making & non-truth-functional. Of course, one has to say what 'true' and 'false' mean in these differing conventions, and this leads us back to the fundamental problem of the meaning of quantum logical connectives. Ultimately, I think, the meaning of the connectives must be taken as deriving from the preexisting Hilbert space formalism which, since it uses classical logic, makes quantum logic parasitic on classical logic. Here we have Copenhagenism in quantum logic – yet another priority of the classical.

What do proofs in NDQL look like? The answer is: mostly straightforward, but sometimes not.

Here are some typical proofs in NDQL. The first two are trivial and the third and fourth less so.

$\vdash P \rightarrow P$

1	(1) P	A
	(2) $P \rightarrow P$	1,1CP

$\vdash -(P \,\&\, -P)$ [noncontradiction]

1	(1) $P \,\&\, -P$	A
	(2) $-(P \,\&\, -P)$	1,1RAA

$\vdash P \vee -P$ [excluded middle]

1	(1) P	A
	(2) $P \rightarrow P$	1,1CP
	(3) $-P \vee (P \,\&\, P)$	2Df.→
4	(4) $-P$	A
4	(5) $P \vee -P$	4∨I
6	(6) $P \,\&\, P$	A
6	(7) P	6&E
6	(8) $P \vee -P$	7∨I
	(9) $P \vee -P$	3,4,5,6,8∨E

$(P \,\&\, Q) \vee (P \,\&\, R) \vdash P \,\&\, (Q \vee R)$ [partial distributivity]

1	(1) $(P \,\&\, Q) \vee (P \,\&\, R)$	A
2	(2) $P \,\&\, Q$	A
2	(3) P	2&E

$(P \& Q) \vee (P \& R) \vdash P \& (Q \vee R)$		[partial distributivity]
2	(4) Q	2&E
2	(5) $(Q \vee R)$	4 ∨ I
2	(6) $P \& (Q \vee R)$	3,5&I
7	(7) $P \& R$	A
7	(8) P	7&E
7	(9) R	7&E
7	(10) $(Q \vee R)$	9 ∨ I
7	(11) $P \& (Q \vee R)$	8,10&I
1	(12) $P \& (Q \vee R)$	1,2,6,711 ∨ E

Oddly enough, our proof of the law of excluded middle depends upon the definition for $\rightarrow$. The proof of partial distributivity is exactly as it would be in Lemmon's system. At line (2) we assume the first disjunct $P \& Q$ and prove the result *with no extra premises,* and at line (7) we assume the other disjunct $P \& R$ and again prove the result with no extra premises. If you try proving the distributive law, you will find that you have to violate the quantum logical restriction on ∨E.

Despite all this the de Morgan laws naturally hold (orthocomplementation depends on them)

$$P \& Q \dashv\vdash -(-P \vee -Q)$$

and

$$P \vee Q \dashv\vdash -(-P \& -Q).$$

(Proving them is a good exercise.) We use them in the next chapter in the form of *derived rules,* enabling us to substitute de Morgan equivalents anywhere within a wff.

In NDQL we restrict some of the rules of inference of the classical propositional calculus as presented by Lemmon and we add a new definition. Can we justify any of these changes in terms of what we know about quantum mechanics?

First, consider RAA.

Suppose we have two propositions, call them P and Q, asserting that the momentum and the position respectively of a particle are restricted to a finite range. We can suppose that P happens to be true.

We know from the complementarity theorem that their conjunction $P \& Q$ is a quantum logical contradiction. Therefore if P happens to be true, neither Q nor $-Q$ can be true. We also know that for any R

$$P \& Q \models R \& -R$$

(here $\models$ is *semantic* entailment), since $P \& Q$ is a quantum logical contradiction. Had we proved

$$P \& Q \vdash R \& -R$$

(quantum logical) RAA would have enabled us to prove

$$\vdash -(P \& Q).$$

But given that

$$P \& Q \vdash R \& -R$$

we can easily prove, quantum logically, that

$$P, Q \vdash R \& -R. \qquad [**]$$

Using the *classical* RAA, we could then infer

$$P \vdash -Q.$$

This says that whenever P is true, Q is true. But we know that all we can validly infer is that it is not the case that Q is true.

Again, from,[**] classical CP enables us to infer

$$P \vdash Q \rightarrow (R \& -R)$$

but

$$\vdash Q \rightarrow (R \& -R) \text{ implies that } \vdash -Q,$$

because, given a proof of the former sequent, we can construct a proof of the latter using [noncontradiction] and *modus tollens*. Using the transitivity of entailment, which holds in both classical logic and quantum logic (you should prove this!), we again have

$$P \vdash -Q.$$

So we see that quantum logical RAA, and quantum logical CP are both expressions of the complementarity theorem.

The definition of $\rightarrow$ is simply a version of the orthomodular condition on quantum logic. For with *modus ponens,* we have the orthomodular equivalent

$$P \& (-P \vee (P \& Q)) \vdash Q,$$

and it so happens that the conditional in quantum logic behaves strangely, though through the work of Gary Hardegree it is well-known that it behaves like a *counterfactual* conditional.

For example, the quantum logical conditional does not allow strengthening of its antecedent. Thus

$$P \rightarrow Q \vdash (P \rightarrow (R \rightarrow Q))$$

is not valid quantum logically, though it is of course valid classically. It is not valid for classical counterfactual conditionals either. A counterexample is available. It might be that 'if Jane were to go to the party, John would be happy' is true, but 'if Jane were to go to the party, then if Jack were to go to the party, John would be happy' is false. (Jack is John's deadly rival for the favours of Jane.)

To John's happiness, Jack's going to the party is an interfering event. Similarly in quantum logic, if P and R are not compatible but are 'interfering' propositions, the truth value of Q will be 'disturbed' by R (imagine it is Heisenberg talking).

Even more starkly

$$P \vdash (Q \rightarrow P),$$

one of the paradoxes of material implication, is not valid in quantum logic. Let P be the proposition 'spin "up" in the z-direction', and let Q be the proposition 'spin "up" at 45° to the z-direction'. Suppose P is true. The proposition $P \rightarrow Q$ is really $-P \vee (P \,\&\, Q)$. But $P \,\&\, Q$ is a quantum logical contradiction. So $P \rightarrow Q$ is just $-P$, which is not true. A similar example shows that

$$-P \vee Q,\ P \vdash Q$$

is not valid in quantum logic. (That is the reason why we have to go to $-P \vee (P \,\&\, Q)$ for a sensible material conditional.)

For similar reasons,

$$P \rightarrow Q,\ Q \rightarrow R \vdash P \rightarrow R$$

is not valid in quantum logic. But the 'metalinguistic' version

$$\vdash P \rightarrow Q \text{ and } \vdash Q \rightarrow R \text{ implies } \vdash P \rightarrow R$$

is valid. Entailment is transitive in quantum logic (Thank Heavens!) even if the material conditional is not.

So much for NDQL. Talk of 'entailment' in quantum logic brings us to that sort of logical system which symmetrizes the notion between premises and conclusion, namely a sequent calculus.

A sequent calculus for quantum logic can be symmetrical in its treatments of & and $\vee$. Compared with a natural deduction system, a sequent calculus can be made mathematically elegant. Some philosophers,[9] captivated by their elegance and by the idea that the logical connectives are defined by the rules they obey in a sequent calculus, have sought a sequent calculus for quantum logic which lays bare the meanings of the connectives.

The idea that the meanings of the connectives are to be found in rules is a contemporary application of the idea 'don't look for the meaning, look for the use'. Unfortunately, it doesn't work for quantum logic because one of the special rules of a sequent calculus – the 'cut' rule – is not eliminable, as it is in classical logic. Because the cut rule is not cuttable, all the derivations of a given sequent might require you to employ a formula (somewhere in the sequents in the proof) which contains a connective which does *not* occur in the final sequent. If the cut rule were cuttable, this would not be the case.

So, what the rules in a sequent calculus for quantum logic say about the connectives separately cannot determine which sequents containing them are and are not provable. Which is to say that the meanings of ∨ and & will depend, in quantum logic, on what – means, an unsatisfactory state of affairs and one which rules out identifying rules with definitions.

We shall see that what the connectives mean is a real problem in the philosophy of quantum mechanics. All attractive routes for defining them independently of the formalism of quantum mechanics seem to be blocked (I think they are blocked). If this is so, the scope of quantum logic as a 'logic of the world' will be restricted (as I think it is).

Quantificational quantum logic

One last logical topic.

So far we have looked upon quantum logic only as a propositional logic, which makes the system seem very restricted. But really, the formal structure of quantificational quantum logic is already there. As in classical logic, the (quantum) universal quantifier behaves like conjunction and the (quantum) existential quantifier like disjunction. Of the four introduction and elimination rules for & and ∨, only ∨E is nonclassical, so we can expect that of the quantifier rules only 'existential quantifier elimination' will be nonclassical. In fact, in quantum logic it is restricted in exactly the same way ∨E is.

The classical rule (in a Lemmon-like system) is already tedious enough to state: suppose you have proved an existentially quantified proposition $(Ex)Px$ on the basis of some premises Γ. In other words, you have proved

$$\Gamma \vdash (Ex)Px$$

Then given Γ, P is going to be true of some individual though we do not know of which. So let the individual of which P is true be **a,** named by a where a is an *arbitrary* name, strictly speaking neither a variable nor a constant. If, having assumed *Pa,* we can, with a set (possibly empty) of additional premises Γ_1, derive a conclusion *C* which is independent of *a,* then we can infer that

$$\Gamma, \Gamma_1 \vdash C.$$

That is, given both Γ and Γ_1, we can derive *C* because in deriving *C* it does not matter which object **a** *P* was true of. There is a further restriction to be placed on Γ_1, namely that *a* should not occur in Γ_1.

The final quantum logical restriction is different, and more severe. It is that Γ_1 has to be *empty*. The effect of restricting ∨E is, as we recall, to make the distributive law underivable. The effect of the restriction on the existential quantifier is to block the distribution of & over the existential quantifier

$$(Ex)Px \,\&\, (Ey)Qy \vdash (Ex)(Ey)(Px \,\&\, Qy).$$

An attempt at a proof would look like this:

1	(1) $(Ex)Px$ & $(Ey)Qy$	A
1	(2) $(Ex)Px$	1&E
1	(3) $(Ey)Qy$	1&E
4	(4) Pa	A
5	(5) Qb	A
4,5	(6) Pa & Qb	4,5&I
4,5	(7) (Ex) $(Px$ & $Qb)$	6EI

. . . at which point one wants to infer

1,5	(8) (Ex) $(Px$ & $Qb)$	1,2,7EE

eliminating the dependence on premise 4, but the presence of the additional premise 5 blocks this.

This idea plays a role in Putnam's defense of quantum logical realism, one of the ideas we examine in the next chapter. It seems to allow us to say 'the electron has definite position and momentum' without our being forced to say 'the electron has definite position and momentum'.

But the behaviour of & and A on the one hand, and of ∨ and E on the other are so similar that we will find we can discuss all the quantum logical resolutions of the paradoxes at the level of propositional logic. Quantification doesn't really add anything new.

10

Quantum logic: what it can and can't do

What do we expect of a quantum logical interpretation of quantum mechanics worthy of the name? In what sense should it *resolve* the paradoxes? What scope should it have? Should quantum logic, if it is successful in the microworld, replace classical logic in the macroworld and in mathematics?

We distinguish between two kinds of quantum logical interpretations of quantum mechanics.

An example of the first kind, the *activist* kind, will do much more than simply rewrite the assertions that quantum mechanics makes about the quantum world in an unfamiliar notation – in the elementary language of quantum mechanics. It will offer a solution to the major puzzles of quantum mechanics. It will show that the paradoxes of quantum mechanics can be resolved if we do away with classical logic and replace it with quantum logic, at least in our descriptions of quantum phenomena. It may defend a form of quantum-mechanical realism, a sort of quantum logical particle view of quantum systems, protecting it from the paradoxes that classical logic would impose on it.

The second kind of quantum logical interpretation, the passive or *quietist* kind, is less ambitious. A quietist interpretation offers a solution to the paradoxes only in the following sense. It will assert that quantum mechanics, when it describes quantum phenomena in the elementary language, need not generate paradox, that the so-called paradoxes are not even formulable in quantum logic.

Both kinds of quantum logical interpretation should also have something to say about how changing logic can enhance rather than further obscure our understanding of the quantum world. It should, in other words, explain what is involved in changing logic and how the key features of that logic – like the meanings of the logical connectives – are to be understood.

Quantum logical interpretations also differ over the scope of quantum logic. In one view of quantum logic – in a *revisionist* view, appropriate to an activist motivation – quantum logic is meant to *replace* classical logic everywhere. Quantum logic is supposed to be the real logic of the macroworld as well as of the microworld and only a mathematics which respects the limitations of the quantum world is a true mathematics. Few philosophers are quite so boldly revisionist.

More common among philosophers of physics is *preservationism*. According to a preservationist account, it is classical logic that reigns in mathematics and in the world of middle-sized objects which we inhabit, although quantum logic is the logic of the microworld. Preservationism is appropriate both to quietist and to activist quantum logical interpretations. But preservationism, while more moderate and reasonable than revisionism, has the problem of how to account for the logical 'cut' between the micro- and macroworlds. There is nothing like the correspondence principle in logic.

This is a point worth emphasizing. Logic cannot 'go over into distributivity' in the limit of large quantum numbers. The failure of the distributive law in quantum logic does not depend on size, it depends only on Planck's constant being nonzero, and between zero and nonzero there is an absolute discontinuity.

An obvious and immediate battery of problems which faces the revisionist quantum logical view are these: How can we explain away the fact that the mathematical apparatus of quantum mechanics, the formalism, is based on the classical logic which is responsible for the paradoxes? How can we give quantum logic a life of its own, independent of the formalism of quantum mechanics – independent of the structure of Hilbert space – whose logic is classical? How can we give *meanings* to the quantum logical connectives without presupposing classical logic?

But these interesting and difficult questions get whatever urgency they have in the philosophy of quantum mechanics from whatever success revisionist quantum logical interpretations have in sorting out the paradoxes of quantum mechanics. If quantum logic fails in its task of sorting out quantum mechanics, then questions about the meanings of the quantum logical connectives revert to being mere puzzles in philosophical logic, and cease to be important in the philosophy of physics.

It is therefore wise to consider first whether quantum logic can do its work. One cannot hope to survey all the possibilities. But one proposal stands out from all the others – that of Hilary Putnam in his brilliant and provocative essay *The Logic of Quantum Mechanics*,[1] the source of much of the interest that philosophers of physics have displayed in quantum logic during the last decade or so. One should say that the Putnam of that paper and the Putnam referred to throughout this chapter is now an historical figure in contemporary philosophy (that historical Putnam seemed to have built-in obsolescence), and has been superseded by newer Putnams. But Putnam put down the prototype (as he so often does) for a particularly radical philosophical idea.

Putnam's account of quantum logic is naturally activist. I shall argue that it fails. I mean to suggest that all similar quantum logical interpretations will fail. However, I shall also sketch what I take to be the justified scope of quantum logic, as the logic of the descriptions of quantum phenomena at the level of the ELQM. Quantum logic does not *resolve,* but rather *embodies* all the strange features of quantum mechanics. The quantum logician who is happy with quantum logic can adopt a quietist pose in the face of the 'paradoxes'. This may not seem much, but it is the best quantum logic has to offer.

Hilary Putnam's logic of quantum mechanics

Putnam's original revisionist quantum logical interpretation of quantum mechanics is extremist in several different ways. However its failure and the reasons why it fails are of interest in part because it illuminates the scope for other quantum logical accounts of quantum mechanics.

First, what were Putnam's motives in proposing a quantum logical interpretation of quantum mechanics? In claiming that the logic of quantum mechanics is nonclassical Putnam's target belonged as much to the philosophy of logic as to the philosophy of physics. He meant to use the proposal of a quantum logical interpretation of quantum mechanics to *illustrate* a reason we might have for throwing over classical logic. If we could have such a reason – an empirical reason – then logic would have to be taken to be an *empirical science*.

As *evidence* for the empirical nature of logic Putnam argued that his quantum logical interpretation should be adopted because it is to be *preferred* to its rivals. It is to be preferred because it is realist without admitting hidden variables, because it is an individual system interpretation, because it requires no 'cut' between the quantum and classical worlds, and because it is paradox-free.

So Putnam *begins* from an empiricist view of logic, since he holds that in choosing a logic we are to make a choice similar to that made by a cosmologist when he chooses a geometry for space-time. One's reason for making the choice that one does is partly pragmatic; one chooses the best logic given a variety of desiderata, one of which, in the case of the logic of quantum mechanics, is quantum-mechanical realism.

There is, Putnam holds, a close, almost perfect analogy between logic and geometry. Thus, it was a discovery of the nineteenth century that geometries other than Euclidean geometry were consistent. Non-Euclidean geometries *are* geometries because they share with Euclidean geometries their defined terms, like 'line' and 'triangle', and because they possess the same axioms, except for the famous axiom of parallels. Similarly, says Putnam, quantum logic is logic because the meanings of the connectives in both classical and quantum logics are identical and because the two logics share a significant body of valid arguments or provable sequents excepting (along with some others) all those that imply distributivity. In Putnam's view distributivity in logic plays a key role, analogous to that of the axiom of the parallels in geometry.

Nowadays we view the geometry of space as a matter to be decided by physics. It is time, so Putnam says, to admit that the *logic of the world* is also a matter for physics. There is a philosophical corollary to this idea: the nineteenth and twentieth centuries have both made inroads into our conception of Necessity, the nineteenth by eliminating the Necessity of Euclidean geometry and the twentieth by doing something even deeper, eliminating the Necessity of some of the theorems of classical logic.

As a philosopher of quantum mechanics the historical (1968) Putnam's main theses are these.

First, the logic of quantum mechanics is non-Boolean.[2]

Second, the peculiarities of quantum mechanics arise through illegitimate uses of classical logic in the description of individual quantum systems.[3] In particular, all cases of *complementarity* correspond to logical incompatibilities in quantum logic;[4] the paradox of the two-slit experiment arises through a quantum logically unacceptable use of the distributive law;[5] and Heisenberg's paradox of the electron in the hydrogen atom does not arise in a quantum logical interpretation of quantum mechanics.[6]

Third, probability in quantum mechanics is classical, only the logic of quantum mechanics is nonclassical.[7]

Fourth, quantum logic licences quantum-mechanical realism.[8]

Fifth, ideal measurements *reveal* values for dynamical variables *possessed* by the measured system prior to measurement.[9]

Sixth, though quantum-mechanical states are not classically complete, they do correspond to quantum logically maximally consistent sets of sentences. Indeterminacy arises not because the laws of quantum mechanics are indeterministic but because quantum-mechanical states are not *classically* complete.[10]

Seventh, the meanings of the quantum logical connectives are the same as those of the classical connectives.[11] Quantum and classical logics are rival logical systems, that is, rival empirical theories about what is analytically true of the logical connectives.

Finally, the success of quantum logic shows that logic is an empirical science like empirical geometry.[12]

Putnam's reasoning depends, as we noted, on an interesting blend of pragmatism and realism of, on the one hand the 'Two Dogmas' Quinean idea that the theorems (Putnam would say: not *all* the theorems) of classical logic are not analytically true, and on the other hand the un-Quinean notion that there are nevertheless analytic truths among which are *some* of the laws of classical logic.

It is worth asking how it is that Putnam can take this line. First, he takes Quine's attack on the analytic–synthetic distinction to be an attack on the *a priori,* rather than on analyticity. That the distinctions between the analytic and the synthetic on the one hand, and the a priori and the a posteriori on the other are the same, was one of the dogmas of logical positivism, and a rejection of the Kantian synthetic a priori. It is Putnam's view that Quine, who took himself to be attacking analyticity, was really attacking the a priori, showing that he, Quine, had failed to emancipate himself entirely from the influence of the 'empiricism' (that is, logical positivism) which he was attacking.

Putnam's view is that we can no longer defend *classical* logic on the grounds of its obvious analyticity, truth in virtue of meaning alone. But some statements *are* analytic and others not. Some laws of some logics – all laws of the true logic – are analytic and some not, but we *discover* which are and which not by *empirical* investigation. Our empirical conclusions are not *forced* on us. Our acceptance of something as 'discovered' has a pragmatic dimension. Putnam claims that what we discover in the philosophy of quantum mechanics is that all the theorems of quantum logic are analytic while some of the theorems of classical logic are not. Or, more briefly, that quantum logic is both analytic and is the logic of the world, and that the fragment of classical logic which is not part of quantum logic is not analytic and is in fact false of the world.[13]

Now we can reconcile a belief in analyticity and a belief that there is an empirically discovered logic of the world. Analyticity is epistemically contingent. Analyticity is not a priori. But the logic of the world is analytic, and so far we have got by with a logic which is approximately analytic, just as we usually get by with a geometry that is only approximately true.

All this is very exciting, not to say strange. However, as far as quantum mechanics goes, it is Putnam's fourth thesis, quantum logical realism, that is his central and dominating idea. Quantum logical realism asserts that (in some peculiar quantum logical way) the values of all the dynamical variables associated with any system are all simultaneously precise. Quantum logical realism governs Putnam's resolution of the paradoxes in general and his idea that probability in quantum mechanics is classical in particular. For us therefore the most fundamental question is this: do Putnam's quantum logical realism and his resolutions of the paradoxes work? The answer is that they don't. We take his resolution of the paradox of the two-slit experiment as archetypical of both failures.

Putnam on the two-slit experiment

Putnam's most thoroughly worked out resolution of the paradoxes is that of the two-slit experiment. He formulates the paradox as a puzzle in probability theory in the following way.

Let $P_{\text{cond}}\,[a|b]$ be the *conditional probability* that a is the case given that b is the case. Write the probability of a as $P[a]$, etc. Then classically

$$P_{\text{cond}}[a|b] = P[a \,\&\, b] | P[b], \quad P[b]>0.$$

The *paradox* of the two-slit experiment takes the following equivalent forms.

(1) Why is the probability $P_{\text{cond}}[r|a_1 \vee a_2]$ – the probability that a photon hits the region of the screen named in r having negotiated the diaphragm with the two slits 1 and 2 open – *not* equal to one half the sum of $P_{\text{cond}}[r|a_1]$ (that it hits the region named in r after going through slit 1) and $P_{\text{cond}}[r|a_2]$ (ditto slit 2)?

(2) Why does the classical result

$$P_{\text{cond}}[r|a_1 \vee a_2] = \tfrac{1}{2}(P._{\text{cond}}[r|a_1] + P_{\text{cond}}[r|a_2]) \qquad \text{[ADD]}$$

break down?

(3) Why are the interference patterns due to the two slits not additive?

Putnam's answer is that the derivation of the false and paradoxical result [ADD] employs the distributive law illegitimately and that you can block the paradox by denying the distributive law. Here is his argument.

From the definition of conditional probability one has

$$P_{\text{cond}}[r|a_1 \vee a_2] = P[r \,\&\, (a_1 \vee a_2)] | P[a_1 \vee a_2] \qquad (1)$$

Expanding the right-hand side of (1) using the distributive law one has

$$P_{\text{cond}}[r|a_1 \vee a_2] = P[(a_1 \,\&\, r) \vee (a_2 \,\&\, r)]|P[a_1 \vee a_2] \qquad (2)$$

$$= P[a_1 \,\&\, r]|P[a_1 \vee a_2] + P[a_2 \,\&\, r]|P[a_1 \vee a_2] \qquad (3)$$

since a_1 & r and a_2 & r are orthogonal and hence compatible.

However $P[a_1] = P[a_2]$ since there is no more reason for the photon to go through one slit than through the other. Furthermore,

$$P[a_1 \vee a_2] = 2P[a_1] = 2P[a_2]$$

since a_1 and a_2 are orthogonal and hence compatible. Thus

$$P[a_1 \,\&\, r]|P[a_1 \vee a_2] = P[a_1 \,\&\, r]|2P[a_1] = P[r|a_1]|2. \qquad (4)$$

Similarly

$$P[a_2 \,\&\, r]|P[a_1 \vee a_2] = P[r|a_2]|2. \qquad (5)$$

Therefore

$$P[r|a_1 \vee a_2] = 1/2(P[r|a_1] + P[r|a_2]). \qquad (6) \quad [\text{ADD}]$$

But we know [ADD] is false. So how can we avoid it?

Putnam's solution is that prevention is better than cure, and so he adopts the superficially convincing strategy of barring the move from (1) to (2), the move which uses the distributive law which, as we know, cannot be relied on in quantum logic. If the derivation can be blocked at this point and if there is no other quantum logically acceptable way in which [ADD] can be derived, then the paradox dissolves. Dissolution is the aim of Putnam's game. (Notice the rather limited and 'unphysical' spirit of this use of quantum logic. Putnam's problem is not to get the *right* answer for the two-slit pattern, which is the physicist's problem. Putnam's problem is to *stop getting the wrong answer*. Avoidance of paradox is, I am afraid, typically the philosopher's but not the physicist's strategy.)

So does Putnam's preventative resolution work?

Notice first that the distributive law holds in one direction even in quantum logic. In NDQL one can (and we did) prove the sequent

$$(a \,\&\, b) \vee (a \,\&\, c) \vdash a \,\&\, (b \vee c).$$

Therefore if quantum probability preserves the logical ordering of propositions – if 'a implies b' entails that $P[a] \leqslant P[b]$ – then one can replace the equality of Putnam's line (2) with '$\leqslant$'. Following through one has at line (6)

$$P_{\mathrm{cond}}[r|a_1 \vee a_2] \geqslant 1/2(P[r|a_1] + P[r|a_2]). \qquad \text{[ADD–QL]}$$

But this is also violated by the two-slit experiment. When *destructive interference* occurs the left-hand side of [ADD–QL] can be less than the right-hand side. In other words one cannot block the paradox, as Putnam tries to do, merely by appealing to a failure of the distributive law.

This is not the only objection to Putnam's quantum logical resolution of the two-slit paradox, but it is by far the most telling because it does not depend on the semantics we associate with quantum logic, and because it does not depend on what we mean by saying that the photon goes through slit 1 or through slit 2.

As a matter of fact it can be argued that the distributive law does not fail at all in the two-slit experiment.[14] Assume that 'the photon goes through slit 1' means 'at some time t earlier than the time of impact on the screen the photon has its state-vector localized to the neighborhood of slit 1', and similarly for slit 2. From the complementarity theorem it follows that the conjunction

$$r \;\&\; (a_1 \vee a_2)$$

is quantum logically false and hence so is its distributive expansion. Therefore the distributive law holds for this case, both sides being quantum logically equivalent, both sides being quantum logical contradictions.

This is a good but not a conclusive reason for doubting the failure of the distributive law in the two-slit experiment. It is not conclusive because it is not entirely clear in the quantum logical account what the photon's going through one slit rather than the other is supposed to mean. The telling objection is that Putnam's resolution of the paradoxes would not work even if the distributive law really does fail.

If the electron is localized at one of the slits and then at a region of the screen, the complementarity theorem tells us that the distributive law does *not* fail. The complementarity theorem also tells us something else, which is worth elaborating on. It tells us not to confuse the failure of distributivity in quantum logic with the indeterminacy relations. In fact, it raises the question of how we are to interpret the indeterminacy relations *on a quantum logical view* of quantum mechanics. There is a good deal of confusion between nondistributivity and indeterminacy. Typical is Edward MacKinnon who writes

Because of the indeterminacy principle, standard logic, which is distributive, does not hold. Consider a particle with momentum p confined in a box of volume V, which may be conceptually subdivided into n subvolumes v_i of such size that $pv_i = h|10$, where h is Planck's constant. Through the use of projection operators one could also let p (or v_i) stand for the statement 'The particle has momentum p (or is in the volume v_i)'. Then the distributive law proper to classical logic suggests:

$$p \,\&\, V = (p \,\&\, v_1) \vee (p \,\&\, v_2) \vee \cdots \vee (p \,\&\, v_n).$$

The left side of this supposed identity is simply a restatement of the boundary conditions. Each disjunct on the right is forbidden by Heisenberg's indeterminacy principle. With the standard definition of '&' and '∨' a true statement comes out logically equivalent to a disjunction of statements each of which is false.[15]

If p and V are each bounded by a finite range, both sides of the identity are quantum logical contradictions and so the distributive law holds, contrary to what MacKinnon says, and the δP_x and δX in the indeterminacy relations do not refer to total spreads but to standard deviations (of measurement results presumably).

The best and simplest thing to say here is that nondistributivity and indeterminacy are not related in any simple way at all. On a simple-minded reading of the truth conditions of ELQM that a proposition is true of a quantum system iff its state-vector belongs to the corresponding subspace – it seems that we must interpret the indeterminacy relations as *statistical* scatter relations. This is of course not to say that we must do so on all other interpretations of quantum mechanics.

To return to Putnam and the two-slit experiment. Clearly one must look deeper for the impact of quantum logic on our understanding of the two-slit experiment than Putnam does. What does quantum logic really tell us about the two-slit experiment?

One oddity of Putnam's treatment of the two-slit experiment lies in his use of the classical probability calculus. The laws of logic may be giving way under us, and under probability as well which is after all a measure on the propositions in ELQM, but classical conditional probability is assumed to survive unscathed. But change the laws of logic, change the structure of the lattice on which probability is a measure and presumably you must change the laws of probability. One must ask: is there a sensible conditional probability in the quantum probability calculus? There is every reason to think that there isn't. Take two complementary propositions p and q. Since $p \,\&\, q$ is a quantum logical contradiction the conditional probability

$$P_{\text{cond}}[p|q] = P[p \,\&\, q] | P[q], \; P[q] > 0 = 0.$$

One question one might ask is: could we substitute a different lattice polynomial – an expression built up from the lattice operations and variables ranging over the lattice elements – for $p \,\&\, q$ in the numerator of the definition to get the classical result for compatible p and q, and an always reasonable generally nonzero otherwise? The answer is no. It is one thing to have a conditionalization rule, like the Luders rule, which tells what the probabilities of the results of a measurement of an observable would be, given that the system is in such and such a state. It is quite another to abstract from this all mention of measurement, and obtain a conditional probability for possession by a system of a value for an observable, given that it possesses a definite value for some incompatible observable.

Therefore quantum logic rules out the use of the classical conditional probability for incompatible propositions. This should have been how Putnam used quantum logic in the two-slit experiment: quantum logic stops you formulating the paradox at all, in the way that Putnam does, as a paradox about probability. However the two-slit experiment is paradoxical only for light beams of quantum systems. The individual quantum system always behaves like a particle. The correct quantum logical resolution of the paradox is that it is not formulable. There are no conditional probabilities to get the thing going. Quantum logical quietism is something we return to when we consider the Bell inequalities against a quantum logical perspective.

Putnam's quantum logical realism

In Putnam's account of the two-slit experiment, the electron goes through one slit or other, but not both. Classical probability reigns because we are dealing with (quantum logical) 'particles'. Clearly, quantum-mechanical realism is motivating and pulling the strings of Putnam's philosophy of quantum mechanics. However it is a peculiar kind of realism, a quantum *logical* realism.

Putnam's resolution of the paradox of the two-slit experiment would have been a remarkable triumph, had it been successful. Quantum logic is the logic of the language in which we describe quantum systems, and an appeal to a failure of a classical law of logic in an attempt to evade a classical description of a quantum particle is quite reasonable. It would be, to speak loosely, an object-language level use of quantum logic.

What sort of argument is there in favour of quantum logical realism, that which motivates Putnam's failed resolution of the paradoxes? The answer is that it is a *meta*language level use of quantum logic, in which quantum logic is used to *avoid* a conclusion about the object-language ELQM. So what is it?

First, consider an example.

Consider the conjunction of two propositions, the first an exhaustive disjunction of statements of ELQM each of which assigns a definite value (that is a value restricted to a finite range, or less acceptably, an 'exact' value) of position $X_1 \vee \cdots \vee X_n$, and the second an exhaustive disjunction of statements of ELQM each of which assigns a definite value of momentum $P_1 \vee \cdots \vee P_m$. (There is the slight problem that we assume our language can handle sufficiently large disjunctions and conjunctions.) The conjunction

$$(X_1 \vee \cdots \vee X_n) \,\&\, (P_1 \vee \cdots \vee P_m) \qquad [*]$$

is quantum logically *valid*. It is quantum logically tautologous. However the distributive expansion

$$(X_1 \,\&\, P_1) \vee \cdots \vee (X_n \,\&\, P_m) \qquad [**]$$

is a quantum logical *contradiction*, because each of the disjuncts is. Therefore, Putnam reasons, realism – the assertion that a quantum system has one of the exact X_i's and one of the exact P_j's – holds because it corresponds to the former conjunction. The mistaken 'classical logical' conception of realism corresponds to the second 'disjunction', the distributive expansion, which is of course false (quantum logically).

The odd thing about this is that it should be thought the assertion of [*], which may be stated as

'the particle has a momentum and it has a position',

is an assertion of realism, whereas [**], which may be stated as

'the particle has a joint position and momentum'

is a realism, but is *too strong* a realism. If asserting [*] is asserting realism, it is not to assert that a quantum system has a trajectory, for that would have to be stated in terms of [**]. Nor would it allow us to say that the correlated electrons in EPR have separate determinate values for their dynamical variables. Quantum logical realism is therefore a realism in name only.

One of the difficulties in reading Putnam's account of quantum mechanics is that he does not give a semantics to his use of quantum logic. This is why, as we noted previously, giving a purely formal counterexample to his resolution of the paradox of the two-slit experiment is better than considering whether or not the distributive law fails. With no semantics we have no good reason for supposing that it does. Friedman and Glymour have considered what sort of semantics Putnam might want to give to quantum logic, given his realism, and they have found that none can be given,[16] a result that should not be surprising. In fact

Putnam was later led to assume that a quantum system has many state-vectors, in fact one for each (nondegenerate) observable.[17] He claimed that though a system had many such state-vectors, we, as users of quantum mechanics, could never *assign* more than one at a time. But these clever moves take us so far away from quantum mechanics proper as to be difficult to pursue, and in fact Putnam didn't.

What do the quantum logical connectives mean?

There may be some use for quantum logic – I shall argue later that there is. If there is, then we should ask: what is it to 'change our logic'? What do we demand of a system that it be a logic, and that it be intelligible to us?

Measured against our classical intuitions, quantum mechanics is a pretty odd theory. Quantum logic inherits that oddness. A quantum logical interpretation of quantum mechanics shifts some of the oddness of quantum mechanics into logic, and so the discussion in the philosophy of logic that quantum logic inspires is bound to look pretty odd itself, the central question inevitably being what is involved in adopting a logic as odd as quantum logic.

Though the success or failure of quantum logical resolutions of the paradoxes is the central issue concerning quantum logic in the philosophy of quantum mechanics, it does not impinge directly on what is for Putnam the main thesis of his empiricist philosophy of logic, which is: can physics force a change in logic?

So we ask not *does,* but *could* adopting quantum logic as our 'logic of the physical world' enhance our understanding of the physical world. One objection that it could not seeks to trivialize it: adopting quantum logic is simply changing the meanings of the connectives 'and', 'or' and 'not'; quantum logic is the result of mapping quantum mechanics onto a language – ELQM – with a bizarre underlying logic.

The *conventionalist* form of this objection adds that the new meanings are merely new conventions and that nothing has really changed. Conventionalism is of course a quite familiar theme in the philosophy of science. General relativity tells us that the geometry of space-time is non-Euclidean, but the conventionalist says that this is simply to change the meaning of the expression 'straight line'. The discoveries that Euclidean geometry not only is not necessarily, but also is not in fact, the geometry of our world are trivialized and seen to be about our language. They are to be seen as conventions we adopt in talking about the world, and not about what we talk about, namely the world itself. In addition the conventionalist would think that the phrase 'the logic of the world' is plain nonsense, the result of what Ryle called a 'category mistake', applying to the world what should be applied to our talk about the world.

One should also demand an account of the meanings of the quantum logical connectives because without one it would not be clear what a quantum logical interpretation involves. One can ask *why* the distributive law fails in quantum logic just as one can ask why the law of excluded middle fails in intuitionistic logic. The answer to both questions will turn on what we mean by the connectives in the two logics. Merely pointing out the *formal* changes that quantum logic involves would not be enough.

Putnam (or at least the Putnam of our period) resists conventionalism in the philosophy of quantum mechanics as he does in the philosophy of space and time. He takes it head on, claiming not only that adopting quantum logic is not *simply* changing the meanings of the connectives, but that it does not involve connectives with changed meanings *at all*. The quantum logical connectives are, in Putnam's view, *synonymous* with the classical connectives. They are essentially the same things. And this claim also answers the independent demand for the meanings of the connectives. Putnam claims this because he thinks[18]

> a strong case could be made out for the view that adopting quantum logic is *not* changing the meanings of the connectives, but merely changing our minds about the law
>
> $$p \wedge (q \vee r) \text{ is equivalent to } (p \wedge q) \vee (p \wedge r).$$

The strong case is this. The following principles (1) to (9) hold both in classical and quantum logics and seem to capture basic features of any 'and', 'or' and (less obviously) 'not':[19]

(1) p implies $p \vee q$
(2) q implies $p \vee q$
(3) if p implies r, and q implies r, then $p \vee q$ implies r
(4) p and q together imply $p \wedge q$
(5) $p \wedge q$ implies p
(6) $p \wedge q$ implies q
(7) p and $-p$ never both hold
(8) $p \vee -p$ holds
(9) $- -p$ is equivalent to p

The respective quantum logical connectives have the *essential* properties of conjunction, disjunction, and negation and these essential properties, in Putnam's view, are sufficient to characterize meanings. The principles (1) to (9), together with their substitution instances, are to be regarded as analytic. The distributive law, which presumably should be seen as a claim about the relationship between 'and' and 'or', and not as a claim about 'and' and 'or' separately, is not analytic. One can then represent the discovery that the logic of the microphysical world is quantal, as entailing the more specific discovery that the distributive law is not analytic, as classical logic takes it to be.

However, can the corresponding quantum logical and classical connectives be synonymous? We could try comparing the corresponding truth tables, if it were agreed what these should be in the case of quantum logic, which it is not. It is clear from what Putnam says that he does not take the truth tables for the connectives to be definitions, so much as empirical claims about them. But if we do compare truth tables we can say that, given reasonable assumptions, the quantum logical connectives cannot be truth functional, whereas the classical connectives of course are.

One might, in opposition to Putnam, make the weaker claim that if the corresponding classical and quantum logical connectives are synonymous, then if one (say classical disjunction) is truth functional, so should the other (quantum logical disjunction) be. Such a claim does not require that the truth tables be the same, though it would be odd if they were not. But there is an elegant argument, due to Geoffrey Hellman,[20] which uses the weakest possible assumptions and which shows that quantum logical disjunction and negation cannot both be truth functional and two-valued. The assumptions are

(1) For any subspaces, M, N if $M \subseteq N$ and M is assigned 'T', then N is also assigned 'T';
(2) For any subspace M, M is assigned 'T' iff $M^{\perp}$ is assigned 'F'.

Hellman's argument amounts to a proof that on any semantics consistent with (1) and (2), some quantum logical disjunctions are true and others false when both their disjuncts are false. Therefore quantum logical disjunction (at least) is non-truth-functional. Hellman's conclusion applies also to quantum logical conjunction, since we have

$$p \,\&\, q = -(-p \vee -q)$$

and so if negation is truth functional, and disjunction isn't, then conjunction isn't.

In Chapter 9 we chose to represent quantum logical conjunction as truth functional, and quantum logical disjunction and negation as non-truth-functional. We gave ELQM a semantics which made a sentence true iff the appropriate state-vector is a member of the Hilbert space, and false otherwise. This convention leads to the following truth tables:

P	Q	$P \quad Q$	$P \vee Q$	$-P$
T	T	T	T	T
T	F	F	T	?
F	T	F	T	
F	F	F	?	

If one adopts the usual alternative convention, which is the one Hellman's arguments points to, that a sentence of ELQM is true iff the appropriate state-vector belongs to the appropriate subspace, false iff it belongs to the orthogonal subspace, and neither true or false otherwise, then one can make negation look rather more truth functional at the expense of the truth functionality of conjunction. No convention can make *all* the connectives simultaneously truth functional,[21] a fact which is equivalent to a 'no-hidden-variables theorem'. So much for comparing truth tables.

As we noted, Putnam does not look to the truth tables for the meanings of the connectives. So how can he hold that two connectives, the quantum and classical disjunctions for example, can be synonymous if their extensions, their truth tables, are different?

One answer would seem to be, as we suggested, that *if* logic is empirical, then the classical truth tables should be seen as a particular *empirical theory* about the connectives, the essential properties of the connectives being given in another way.

But why should just these properties – Putnam's (1) to (9) – of the connectives, those that make &, v, and – correspond to the operations of *meet, join,* and *orthocomplementation* in an ortholattice, be the essential properties which define the connectives? Putnam's claim seems to be simply arbitrary. What reason is there for holding that these properties, and not those expressed in the truth tables, are the essential of properties of classical negation, conjunction, and disjunction? Is Putnam's (3), one of the lattice conditions for 'join', really essential for any disjunction? How would you convince someone it is?

In a paper by Bell and Hallett[22] we find the interesting claim that & and v are synonymous with their classical counterparts, but, contra Putnam, that ' – ' in quantum logic *deviates* in meaning (proving that once you get into the meaning game all combinations of ideas are possible). Bell and Hallett hold this because the binary connectives, viewed as lattice operations, are definable by means of the same formulae in both quantum and classical logics. They hold that this is not so for ' – '. They argue that if

> one takes an extensional view of meaning by stating that two terms have the same meaning relative to **a, b,** . . . if they have *equivalent* definitions in terms

of **a, b,** . . . then a case can be made out that the meanings of the connectives do *not* vary in the passage from classical to quantum logic.[23]

Thus, the 'join' $\vee$ in a lattice is defined by

$$a \vee b = c \text{ iff } \{x : c \leqslant x\} = \{x : a \leqslant x\} \cup \{x : b \leqslant x\}.$$

A dual definition goes for the 'meet', &. However, in a Boolean algebra the orthocomplementation $\perp$ can be defined by

$$a = b^{\perp} \text{ iff } \{x : x \leqslant a\} = \{x : b \ \& \ x = 0\},^{24}$$

whereas in quantum logic 'negation' cannot be so defined.

So are disjunction and conjunction in quantum logic synonymous with their corresponding connectives in classical logic? Only if, on their account, all Bell and Hallett's **a**s and **b**s are the same **a**s and **b**s for both classical and quantal definitions, and they are not, since the partial ordering ('entailment') relation $\leqslant$ is not fixed. Relative to a fixed $\leqslant$, it may be that a case for synonymy could be made out. But $\leqslant$ have different extensions in the two lattices.

One might reply to this point that $\leqslant$ is fixed in spite of the appearance otherwise. It is defined as a reflexive, antisymmetric and transitive relation in both. But why should one pick out these properties as the essential properties of 'entailment', and think of the rest as accidental? We are back to the arbitrariness (and nonextensionality surely) of Putnam.

What Bell and Hallett offer is in fact a variant of Putnam's synonymy claim. They hold that all joins are synonymous and all meets are synonymous because each is a least upper bound or greatest lower bound. Presumably also that all partial orders are synonymous because they all enjoy the same three essential logical properties. The argument against this is again its arbitrariness.

Abandoning these rationalist adventures, we find that there is, in Putnam's abundant paper, a second account of the meanings of the connectives, one which is verificationist rather than essentialist in spirit. The meanings of the connectives, Putnam says, are given by the 'logic of tests' performed in quantum-mechanical experiments. Of course, operational 'definitions' are not to be taken seriously as definitions. Few philosophers of science believe in operationism as a theory of the meanings of theoretical terms. They are, however, useful as heuristic devices which can give one a better grasp of the theory[25] and the teaching of real physics is full of examples of them.

So, armed with the usual caveats about idealization, we:

> pretend that to every physical property *P* there corresponds a test *T* such that something has *P* just in case it 'passes' *T* (i.e. it *would* pass *T*, if *T* were performed).[26]

The inclusion relation $T_1 \langle T_2$ – which holds just in case everything that 'passes' T_1 'passes' T_2 – generates a partial ordering on equivalence classes of tests, these being classes of tests which pass exactly the same things. There is an 'impossible test', passed by nothing, and a 'vacuous test' passed by everything. Putnam's problem is then to generate an orthocomplemented lattice from this posit of equivalence classes of tests.

Let us speak ambiguously of properties P, and propositions 'P' that assert that a system has P. So let $T_{P \vee Q}$ be the test that corresponds to '$P \vee Q$'. Since everything that has P and everything that has Q should pass the test $T_{P \vee Q}$ we have

$$T_P < T_{P \vee Q} \text{ and } T_Q < T_{P \vee Q}$$

Since any test which passes all things with P and all things with Q must be an upper bound of T_P and T_Q Putnam asserts that

> if there is *any* test at all (even 'idealising', as we have been) which corresponds to the *disjunction* $P \vee Q$, it must have the property of being a *least upper bound* on T_P and T_Q.[27]

One might object that we have no independent reason to assert that there is a test corresponding to the disjunction '$P \vee Q$' – the *least* upper bound of T_P and T_Q – that we must *stipulate* that there is such a test,[28] and that only the Hilbert space formalism assures us that, in principle, there is. It is the test that corresponds to passing just those systems whose state-vectors lie in the subspace spanned by the subspaces corresponding to P and Q. This point puts the operational definition in its place. It should not be thought capable of defining the connectives, though it goes some way to conveying what the significance of quantum logical disjunction on the macrolevel is.

As a matter of fact, realizing the operational definitions has its problems, problems that are sometimes thought sufficient to rule out as 'meaningless' conjunctions and disjunctions of incompatible propositions. Take the case of conjunction for example. A test for the conjunction P & Q where P and Q are incompatible (but perhaps not *totally* incompatible) cannot be a test for P followed by a test for Q since there are interfering measurements. In fact, the test for conjunction has to be an infinite alternating sequence of tests for P and Q[29] which is hardly 'operational'.

It isn't a bad idea to ask how we actually learn what the quantum logical connectives mean. I think we should say that the connectives get their meanings from their connection with the Hilbert space formalism of quantum mechanics, itself formulated in a mathematics whose logic is classical. Operationism fails, partly because the operations which define the connectives cannot be realized, and partly because the definitions presuppose quantum mechanics. So we are back with the formalism, and quantum logic is parasitic on classical logic. The meanings of the quantum logical connectives derive from the Hilbert space formalism and not the other way around.

A 'preservationist', one who thinks that quantum logic, though an acceptable local reformation, should not be held to force an overall revision in our logic, should not be averse to this idea. For then the dependence of quantum logic on classical logic poses no threat. Quantum logic is the logic of the microphysical world, and to grasp it we must climb up the ladder of classical mathematics and classical logic. Quantum logic is a logic, but of a small chunk of our means of describing the world.

Quantum logical mathematics

What does quantum logic do to mathematics?

One might think that a revisionist quantum logical interpretation of quantum mechanics would be vitiated by the fact that the Hilbert space formalism has a logic which is classical. Against this, the revisionist might urge an instrumentalist account of the formalism: namely, that the formalism enables us to make inferences about and predictions about quantum systems, but that the formalism – those bits of it (if any) which are not quantum logically valid bits of mathematics – corresponds to nothing in the world, and is itself neither true nor false. The facts about the world are captured in a language – ELQM – whose logic is quantal.

This application of instrumentalism is no more implausible than any other, which is not to say that it is at all plausible. One can ask why it is that the formalism of quantum theory, as 'instrument', is so useful. One might add that the metaphor of the 'instrument' or 'tool' really is suggestive: we do ask what makes a tool (say a machine tool) effective, and the answer is that it is a precision instrument, machined to very small tolerances, etc. The instrumentalist has no answer to the question: why does the mathematics work? The realist about the formalism does: its structure reflects the structure of the underlying physical facts, however indirectly. This may not be an entirely pleasing answer but it is less disturbing than the instrumentalist view, according to which the success of the mathematics of quantum theory is little short of miraculous.

So, having resisted the attractions of instrumentalism, it seems that the revisionist should consider what a quantum logical mathematics would look like, and whether quantum theory in particular could be reconstructed on its basis. If the meaning of the quantum logical connectives derives from the formalism of quantum mechanics, then we ought to replace the classical logical base of that formalism by quantum logic.

One has to report that very little work has as yet been done on the problem, partly because of the daunting complexity of quantum logical mathematics. But such work as has been done is not altogether discouraging. Michael Dunn, in an interesting paper[30] has shown that extending quantum logic with appropriate axioms for arithmetic is sufficient to make the new arithmetic plus logic distributive. The distributive law is a product of the arithmetic rather than the underlying logic. On the other hand, Gaisi Takeuti[31] has shown that a quantum logical set theory is nonextensional – and hence quite unlike anything we want to countenance as a 'set' theory – and that imposing extensionality on quantum set theory is equivalent to imposing distributivity and hence classical logic from the start.

On this note of inevitable agnosticism we leave quantum logical mathematics, still sceptical of securing a sensible foundation for revisionist quantum logic.

What can quantum logic do for quantum mechanics?

It is time to be more positive. It would be surprising if quantum logic – a logic, or at least a logic-like formal system, which is induced in the fundamental state-space of quantum mechanics – had nothing to contribute to our understanding of quantum mechanics. Quantum logic, I argue, provides the logic for our description of quantum systems at an intermediate level.

The language whose underlying logic is quantum logic – ELQM – is not a low-level or fine-grained language. It is an inexpressive language. It does not admit probabilities. It allows propositions to be certainly true or certainly false or neither. Alternatively, depending on one's semantics, it allows propositions to be certainly true, or otherwise false.

To describe a system using ELQM and quantum logic is somewhat like using a high-level computer programming language. One loses sight of some of the low-level operations of the underlying machine. But one acquires a new way of looking at the machine. Similarly quantum logic is the logic of our high-level descriptions of quantum systems.

Can quantum logic, viewed thus, resolve the paradoxes?

I claim that coupled with an individual system interpretation of quantum mechanics it can at best sustain quantum logical *quietism*. One should not be surprised at this, since it reflects so intimately the structure of quantum mechanics itself. For quantum mechanics, as formalism, there is no paradox of the two-slit experiment. There is only the two-slit experiment. Similarly, for the quantum logician, quantum logic does not *resolve* the paradox, it prevents its being formulated: the conditional probabilities which get the paradox going are not definable in a nontrivial way on a quantum logic.

The Bell inequalities provide a second example. In quantum logic the Bell inequalities are *not derivable,* as indeed they should not be, since they are not derivable in quantum mechanics. To illustrate this, we consider a familiar, simplified derivation,[32] of a simpler and less directly testable inequality than the [BELL-1] of Chapter 8.

Imagine that we have a local hidden-variables theory which determines how both the two electrons in the correlated pair singlet state respond to a spin measurement in any given direction. Naturally the states of the electrons will, according to the theory, dictate that the electrons respond to spin measurements in a fixed direction with oppositely directed spins.

Consider three directions **a, b,** and **c** chosen at random. In any ensemble of electron pairs an electron will be predetermined, according to the local hidden-variables theory, to respond with either 'up' or 'down' to a spin measurement in each of three directions **a, b,** and **c.** Let the number of electrons in the ensemble predetermined to respond with 'up' in each direction be 'num(+, +, +)'. Let the number responding with 'down' in the **a** and **b** directions and 'up' in the **c** direction be 'num(−, −, +)' etc. etc. Finally, let the number of electrons responding with 'up' in the **a** direction, 'down' in the **b** direction and *either* 'up' *or* 'down' in the **c** direction be 'num(+, −, ?)'. The question-mark signifies 'don't care'.

Clearly if N is the total number of electrons

$$N = \text{num}(?, ?, ?).$$

Equally clearly

$$\text{num}(?, -, +) = \text{num}(+, -, +) + \text{num}(-, -, +) \qquad [\text{I}]$$

$$\text{num}(+, ?, +) = \text{num}(+, +, +) + \text{num}(+, -, +) \qquad [\text{II}]$$

$$\text{num}(+, +, ?) = \text{num}(+, +, +) + \text{num}(+, +, -). \qquad [\text{III}]$$

Adding equations [I] and [III] we have

$$\text{num}(?, -, +) + \text{num}(+, +, ?) =$$
$$\text{num}(+, -, +) + \text{num}(-, -, +) + \text{num}(+, +, +) + \text{num}(+, +, -)$$

which, using [II], gives

$$\text{num}(?, -, +) + \text{num}(+, +, ?) =$$
$$\text{num}(+, ?, +) + \text{num}(-, -, +) + \text{num}(+, +, -) \qquad [\text{IV}]$$

or, throwing away some information

$$\text{num}(?, -, +) + \text{num}(+, +, ?) \geqslant \text{num}(+, ?, +) \quad \text{[BELL–2]}$$

which is one of Bell's inequalities.

These quantities 'num(?, −, +)' etc. are measurable for correlated systems and are predictable by quantum mechanics. We know that both quantum mechanics and experiment violate the inequality.

How can we avoid this conclusion quantum logically? The problem is not to find a way of showing that quantum mechanics really does satisfy the Bell inequalities. It clearly does not. The problem for the quantum logical quietist is to *avoid the formulation* of the Bell inequality.

Notice that equations [I] to [III] implicitly assume that, for three (not necessarily compatible) propositions P, Q, and R it is the case that

$$(P \,\&\, Q) \,\&\, (R \vee -R) \dashv\vdash ((P \,\&\, Q) \,\&\, R) \vee ((P \,\&\, Q) \,\&\, -R)$$

which is not valid in quantum logic, as we know.

However, we do have one half of this, namely

$$((P - Q) \,\&\, R) \vee ((P \,\&\, Q) \,\&\, -R) \vdash (P \,\&\, Q) \,\&\, (R \vee -R)$$

so that, as in the case of the two-slit experiment, equations [I] to [III] become the inequalities

$$\text{num}(?, -, +) \geqslant \text{num}(+, -, +) + \text{num}(-, -, +) \quad [\text{I}']$$

$$\text{num}(+, ?, +) \geqslant \text{num}(+, +, +) + \text{num}(+, -, +) \quad [\text{II}']$$

$$\text{num}(+, +, ?) \geqslant \text{num}(+, +, +) + \text{num}(+, +, -) \quad [\text{III}']$$

which give

$$\text{num}(?, -, +) + \text{num}(+, +, ?) \geqslant \text{num}(+, ?, +) + \text{num}(-, -, +) + \text{num}(+, +, -) \quad [\text{IV}']$$

as before but with $\geqslant$ instead of $=$. But then, with mere *inequalities* we cannot get

$$\text{num}(?, -, +) + \text{num}(+, +, ?) \geqslant \text{num}(+, ?, +). \quad [\text{BELL-2}']$$

So the Bell inequalities are not derivable quantum logically. This is not to say that quantum logic *removes* the paradoxes of nonlocality. Rather quantum logic merely has them built-in, so they are not quantum logical paradoxes.

The ignorance interpretation of mixtures provides a third application for quantum logic.

Recall our example of the density operator

$$\rho = \tfrac{1}{2}|+\rangle\langle+| + \tfrac{1}{2}|-\rangle\langle-|$$

which can be interpreted as describing a mixture of two subensembles, one of electrons whose spin is 'up' and the other of electrons whose spin is 'down', each of the subensembles having equal weighting in the mixture.

In speaking of 'up' and 'down' we did not mention a direction. Indeed it is a property of ρ that we can choose *any* direction and represent it thus. But according to the ignorance interpretation of mixtures each individual system in an ensemble belongs to one and only one pure subensemble in a mixture (of orthogonal subensembles). The eigenvalues of the density operator for the mixture (in our case $\frac{1}{2}$ and $\frac{1}{2}$) express a measure of our knowledge that a chosen individual system belongs to the corresponding pure subensemble.

In other words, the ignorance interpretation of mixtures embodies the classical notion of ensemble, as decomposable into a unique set of homogeneous subensembles. However, representing the density operator ρ as decomposed into two pure subensembles of spin 'up' and 'down' in any direction is perfectly correct as mathematics. The problem is how to interpret the mathematics.

The ignorance interpretation seems to collapse into incoherence. How can an electron have either spin 'up' or spin 'down' in (say) the x-direction, and also in the z-direction? It is just as if we were to say that an urn contained a collection of balls each of which is either black or white, *and* either red or green.

Of course things can be thus in quantum logic. Representing spin 'up' in the x-direction as X_+ etc., we can have

$$(X_+ \vee X_-) \;\&\; (Z_+ \vee Z_-)$$

even though each of the four conjunctions of the distributive expansion is a quantum logical contradiction.

This much is supported by a quietist quantum logical interpretation: there just is no problem with the ignorance interpretation of mixtures. Of course, the quantum logical ignorance interpretation of mixtures does not sustain the idea that there is a privileged decomposition of a mixture. But that was part of what we wanted.

One final application of quantum logic: the role of the quantum logical conditional in the quantum theory of *measurement*. The following example has been discussed in somewhat different terms, and with different conclusions by Geoffrey Hellman[33] and Allen Stairs.[34]

If von Neumann's rule is the 'clumsy experimenter's rule', the Luders rule for the projection postulate is, as we noted, the 'careful experimenter's rule'. The Luders rule describes minimally disturbing measurements. Thus suppose we make a nonmaximal measurement the observable O on a system in state $| \rangle$ which is not an eigenstate of O, and that we get the result that the system terminates with an O-value of 'either O_1 or O_2'. We know that the state-vector of the system has been projected into the subspace of Hilbert space spanned by O_1 and O_2. According to the von Neumann rule the new state-vector for the system could be anywhere in this subspace. But according to the Luders' rule, the state-vector should be projected onto the state in the new subspace 'most similar' to the original state-vector. In other words, the state-vector should be minimally disturbed.

Describe this situation in quantum logical terms. Let the original state correspond to the proposition P. Let Q correspond to the subspace into which the system is thrown. (Q need not correspond to a two-dimensional subspace, as it does in our example. The subspace can be as large as you like.)

The state 'most similar' to the original state clearly lies in the subspace corresponding to Q. But it also lies in the subspace spanned by P and $-Q$. (Think of Q as corresponding to the plane of the floor of the room you are in, of $-Q$ as corresponding to the vertical axis, and of P as corresponding to a vector at 45° to all three axes. The new state is the projection of the original state onto the floor. It is in the plane of the floor, and also in the plane spanned by the original vector and the vertical.)

So we must show quantum logically that given that P is true, if Q were true, then so would be $Q \mathbin{\&} (P \vee -Q)$. In other words, to show that the quantum logical conditional is a counterfactual conditional which tells us what would be the case following a minimally disturbing measurement we must prove, in NDQL, the sequent

$$P \vdash Q \rightarrow (Q \mathbin{\&} (P \vee -Q))$$

We sketch a proof of this sequent in NDQL, assuming a derived rule *De Morgan* which lets us make 'De Morgan' substitutions within a wff, and the derived rule &-simplification, which allows us to substitute

$$Q \mathbin{\&} R$$

for occurrences of

$$Q \mathbin{\&} (Q \mathbin{\&} R).$$

We also assume that we can apply Df. $\rightarrow$ as a substitution rule within a wff.

What does this tell us? Not that the Luders rule is explained, or that

$P \vdash Q \rightarrow (Q \,\&\, (P \vee -Q))$ [Luders' rule in quantum logic]

1	(1) $-(Q \rightarrow (Q \,\&\, (P \vee -Q)))$	A
1	(2) $-(-Q \vee (Q \,\&\, Q \,\&\, (P \vee -Q)))$	1 Df.→
1	(3) $-(-Q \vee (Q \,\&\, (P \vee -Q)))$	2 &-simplification
1	(4) $Q \,\&\, -(Q \,\&\, (P \vee -Q))$	3 De Morgan
1	(5) $Q \,\&\, (-Q \vee -(P \vee -Q))$	4 De Morgan
1	(6) $Q \,\&\, (-Q \vee (-P \,\&\, Q))$	5 De Morgan
1	(7) Q	6&E
1	(8) $-Q \vee (-P \,\&\, Q)$	5&E
1	(9) $Q \rightarrow -P$	8Df.→
1	(10) $-P$	9,7MP
	(11) $-(Q \rightarrow (Q \,\&\, (P \,\&\, -Q))) \rightarrow -P$	1,10CP
11	(12) P	A
11	(13) $--P$	12DN
11	(14) $--(Q \rightarrow (Q \,\&\, (P \,\&\, -Q)))$	11,13MT
11	(15) $Q \rightarrow (Q \,\&\, (P \vee -Q))$	14DN

it is not *ad hoc*. It tells us that the Luders rule is represented in quantum logic by a particularly attractive material conditional which is well-behaved. But the Luders rule is no more ad hoc than our material conditional. In fact, we should think of our choice of projection postulate as being decided by experiment. Von Neumann's rule is well-suited to randomizing experiments, Luders' rule to minimally disturbing ones.

Of course, quantum logic cannot explain *why* there should be a projection, only that if there is one, and if it is minimally disturbing, then it is well-described by a well-behaved quantum logical conditional.

So quantum logic welcomes quantum logical quietism. It is mostly easily coupled to preservationism. But that leaves us with the unresolved problem of measurement in the form of the 'cut' between micro- and macroworlds. There is no way of papering over this, except to say that the world is very complicated, that it has many levels, that life is larger than quantum logic (though smaller in some respects than classical logic). Which is no answer at all.

Finally, of all the more conventional interpretations of quantum mechanics, quietist quantum logical preservationism is closest in spirit to Bohr's Copenhagenism, or at least to that part of it which seeks to limit how far we can describe and explain the world. What we learn from quantum theory is how vain we, as human animals, must have been to think, as we did in the classical period, that we possessed the resources to capture the way the world is so neatly and tidily.

Conclusion

Quantum logic is the logic of the high-level language in which we describe what is true and false of quantum systems. I use the term 'high-level' in its computational sense. The language whose logic quantum logic is, is inexpressive. It has no machinery for describing probabilities, only truths and falsehoods and perhaps whatever is in between. Quantum logic conceals from its user the underlying details of the formalism of quantum mechanics, and the details (whatever they may be) of the underlying 'machine', the physical world. Quantum logic encourages a top-down view of the quantum world. Quantum mechanics, as done by the practising physicist, is bottom-up.

The program of *feeding back* quantum logic into our metalevel discussion of realism fails. Quantum logic does not licence quantum-mechanical realism. It cannot override or rewrite the bottom-up view. The program itself is odd in that quantum logic is most naturally thought of as expressing quantum-mechanical antirealism, just as quantum mechanics itself is most naturally interpreted antirealistically. Thus quantum logic is consonant with nonlocality, in the sense that it does not allow one to derive the Bell inequalities. The conditional in quantum logic nicely expresses the effect of ideal measurements on quantum systems.

Even in a quantum logical interpretation, the quantum world remains like a dream. In fact, literally so. The denizens of my nightly dreams have no biographies in my dreams, other than the events of my dreams. They have no height, weight, or birthday. One is antirealist about one's dreams. Similarly quantum systems lack truth about their properties. For many Ps, neither P nor $-P$ is true. Quantum logic reflects the structure of this dreamlike failure of realism.

But quantum logic is a logic. The usual contrast between quantum logic as, on the one hand, the real logic of the world, and on the other, as a merely logiclike algebraic system, is too coarsely drawn. Quantum logic is the logic of a certain important fragment of our talk about the world, just as (perhaps even more controversially) some modal logic is the logic of our talk about necessity.

Within a quantum logical rendering of what it is that quantum mechanics asserts about the microphysical world, it is difficult to interpret the indeterminancy relations other than statistically. The attempt to apply them to the individual system involves rejecting the usual semantics for quantum logic, that a proposition is true of a quantum system if and only if its state-vector is in the appropriate subspace.

The philosophy of quantum mechanics is partly philosophy and partly physics. It is a difficult subject partly because it is inevitably technical but also because it is a hybrid. It is a complex, labyrinthine and at times obscure and opaque area of the philosophy of science, spanned as it must be by theoretical physics on the one hand and by the philosophies of science, logic, and language on the other.

Both philosophers and physicists have their own special contributions to make to the philosophy of physics in general, and to the philosophy of quantum mechanics in particular. Physicists are not only better at the physics, they also have a truer feel for what is physically reasonable and for what isn't. The philosopher's lack of the physicist's *feel* for the physics can be a strength, freeing him to explore the unreasonable, which is after all what quantum mechanics and quantum logic are.

My own philosophical prejudice is that the philosophy of physics is at its healthiest when it is most closely tied to real physics. What I take to be the failure of quantum logic in providing an easy resolution of the paradoxes is, I feel, just a special case of the failure of philosophical fancy footwork in the philosophy of physics.

Notes

All philosophers of physics owe a debt to Max Jammer for his monumental book *The Philosophy of Quantum Mechanics,* published by John Wiley in 1974. Jammer's book was the first survey of all the major areas of the philosophy of quantum mechanics.

A similar debt is owed to J. A. Wheeler and W. H. Zurek for their edited collection *Quantum Theory and Measurement,* Princeton University Press, 1983, in which you can find most of the classical documents of the early phase of the philosophy of quantum mechanics, including the first readily available translation of Heisenberg's γ-ray microscope paper.

Quantum logicians will also need to refer to the excellent collections edited by C. A. Hooker entitled *The Logico-Algebraic Approach to Quantum Mechanics,* Vols I and II, 1979, Reidel.

Preface

1 Jauch (1968) p. v.

Chapter 1

1 Newton from *The Opticks,* quoted from the (1952) Dover edition. p. 400.
2 Laplace (1814) from the *Essai Philosophique sur les Probabilities* pp. 3–4. The translation is from the Dover edition.
3 Very little philosophical work has as yet been done on the specific contribution of quantum field theory to the philosophy of quantum mechanics. But for a good example of some recent work see M. L. G. Redhead (1982).
4 Collected for example in Cartwright (1983).
5 Meaning Hacking (1983). The different works of both Nancy Cartwright and Ian Hacking represent differing reactions to some of the dominant trends in recent philosophy of physics.

Ian Hacking's *Representing and Intervening* is intended to upgrade the role of experiment at the expense of theory in our image of physics, and also to set out an account of Hacking's activist theory of *representing* as the source of ontological commitment. In Hacking's view, Man is *homo depictor,* by inner compulsion a maker of public representations which have the ontological force only when he uses them and not when he merely contemplates them. The life of physics is larger than logic: there is always something in our representations of the world that logic cannot capture.

A different but related response to orthodoxy consists less in playing up experiment and more in playing down the logical coherence of theorising and the truth of theories in physics: Cartwright's line in her recent book *How the Laws of Physics Lie*. Cartwright thinks of physics as factoidal, as less like logic and as more like drama documentary [Cartwright (1983) pp. 139–40]. In physics we select phenomena, posit entities intended to be real, and weave around them a story which need not be literally true. In particular, Cartwright thinks that we overemphasize *mathematical physics* and confuse the real physics with the network of inferences to which the somewhat arbitrary mathematics of physics is liable to confine us. This idea is one of the props of Cartwright's views on quantum mechanics.

Broadly speaking, an image of physics like Hacking's or like Cartwright's will have it that one must understand physics as a human creation and in the context of its use by people, and that the problems which define the philosophy of physics arise naturally as a part of doing real physics. What physics yields cannot be separated from how mathematical physics is *used* and *learned* and from how *experiments* are actually performed. Metaphysics, a sound philosophy of physics, will be the result of seeing physics in action and not simply of giving the formalisms of physics the benefit of a logical analysis.

Chapter 2

1 For a survey of the old quantum theory see the introduction to van der Waerden (1967) pp. 1–18.
2 McCormmach (1982). For a discussion of de Broglie's work and its influence on Schrödinger see Raman and Forman (1969).
3 The reference is to Roger Stuewer's excellent publication (1975).

Chapter 3

1 See G. Peacock (1855) p. 187ff.
2 G. I. Taylor, in the article 'Interference fringes with feeble light', in Taylor (1971) pp. 1–2.
3 See J. A. Wheeler, 'Law without law', in Wheeler and Zurek (1983) pp. 184–200.
4 For a good, up-to-date history of wave–particle duality see Wheaton (1983).
5 W. Heisenberg in Rozenthal (1967) p. 103.
6 The quotation is from M. Born's article of 1926, translated as 'Quantum mechanics of collision processes', in *Wave Mechanics* by G. Ludwig (1968). It is contained in pp. 207–8.
7 From E. Schrödinger (1935), 'The Present Situation in Quantum Mechanics' (the cat paper), in Wheeler and Zurek (1983) pp. 152–67. The quotation is from page 157.

Chapter 4

Heisenberg's article 'Quantum theory and its interpretation' in the volume *Niels Bohr*, edited by S. Rozenthal (1967), is an excellent source of information on the early development of the Copenhagen interpretation. The quotations labelled 'Heisenberg (1967)' are from that volume.

Heisenberg's classic paper of 1927 'The physical content of quantum kinematics and dynamics', Bohr's Como Lecture of 1927, his reply to Einstein–Podolsky–Rosen of 1935, and Bohr's essay 'Discussion with Einstein on epistemological problems in atomic physics' are essential reading. In these notes those articles are labelled 'Heisenberg (1927)', 'Bohr (1935)' and 'Bohr (1949)' respectively. The page numbers cited are to the reprints of those papers in Wheeler and Zurek (1983).

Of the many essays on Bohr's philosophy of quantum mechanics two are very well worth reading: C. A. Hooker (1972), especially pp. 132–47, and E. Scheibe (1973) Chapter 1, pp. 9–49. Hooker's piece, at two hundred odd pages, is more of a book than an article. Scheibe's chapter doesn't read very well in translation but is worth persisting with.

1 Heisenberg (1967) pp. 106–7.
2 For a stout defense of the more dogmatic kind of Copenhagenism see Rosenfeld's paper 'Misunderstandings about the foundations of quantum theory' in Korner (1957).
3 Heisenberg (1967) p. 95.
4 Heisenberg (1967) p. 98.
5 As described in Heisenberg's essay 'First encounter with the atomic concept' in his *Physics and Beyond* (1971) pp. 2ff.
6 That is, Heisenberg (1927).
7 Heisenberg (1927) quoted from Wheeler and Zurek (1983) p. 64.
8 That is, Heisenberg (1930).
9 Again, as described by Heisenberg in his essay 'Quantum mechanics and a talk with Einstein' in *Physics and Beyond* (1971) pp. 58–69.
10 Heisenberg (1949) p. 20.
11 Bohr (1949), quoted from Wheeler and Zurek (1983) p. 17.
12 Bohr (1927), quoted from Wheeler and Zurek (1983) p. 88.
13 Bohr (1927) op. cit. p. 87.
14 Bohr (1927) op. cit. p. 91.
15 Bohr (1927) op. cit. pp. 89–90.
16 Wittgenstein (1968) Parable 139(*a*).

Chapter 5

1 Bohr (1949) 'Discussion with Einstein on epistemological problems in atomic physics', see Wheeler and Zurek (1983) pp. 9–49.
2 Bohr's remark at the 1921 Solvay Congress in Paris, quoted in McCormmach (1970) p. 19.
3 Reprinted in van der Waerden (1967) 159–76.
4 The EPR paper is reprinted in Wheeler and Zurek (1983). The quotation is from p. 138 of that volume.
5 EPR, op. cit. p. 138.
6 See Bohm (1951), *Quantum Theory,* pp. 611–23. The section is reprinted in Wheeler and Zurek (1983) 356–68. In the original book *Quantum Theory,* it appears in pp. 611ff in the (1961) Prentice–Hall edition.
7 Wu, C. S. and Shaknov, I. (1950).
8 Bohr's reply is reprinted in Wheeler and Zurek (1983) pp. 145–51. The quotation comes from p. 148.
9 F.J. Belinfante (1975) p. 8.
10 Schrödinger (1935), quoted from the translation in Wheeler and Zurek (1983) p. 157.

11 Ballentine (1970).
12 Ballentine (1970) p. 364.
13, 14 Discussed in Ballentine op. cit., p. 362 and Popper (1967) p. 24.
15 See for example Rosa (1979) on electron interference.
16, 17 Popper (1967) pp. 32–3.
18 Popper op. cit.
19 Popper op. cit., p. 21.
20, 21 Popper op. cit., p. 33.
22 Popper op. cit., p. 35.
23 See Feyerabend (1968 & 1969).
24 From L.D. Mermin's review of two of Popper's recent volumes. Mermin (1983) p. 656.

Chapter 6

1 From *Against Method:*

von Neumann replaces the quasi-intuitive notions of Dirac (and Bohr) by more complex notions of his own . . . the theory becomes a veritable monster of rigour and precision while its relation to experience is more obscure than ever.' Feyerabend (1975) p. 64 fn. 23.

2 The terminology for this rather restricted language seems to have been introduced in van Fraassen (1970).

Chapter 7

1 von Neumann (1955) Chapter VI. p. 417ff.
2 Luders' original discussion is to be found in Luders (1951). But see also the discussion of von Neumann's and Luders' postulates especially with respect to the two-slit experiment in Bub (1979).
3 See Wigner (1961), reprinted in Wheeler and Zurek (1983) pp. 168–81.
4 For a good account of this see the essays by Bryce S. DeWitt in DeWitt and Graham (1973) pp. 155–65 and pp. 167–218.
5 An excellent critique of the many-worlds interpretation can be found in Healey (1985).
6 Daneri, Loinger, and Prosperi (1962), reprinted in Wheeler and Zurek (1983) pp. 657–79.
7 A remark in Cartwright (1983), p. 171.
8 The SLAC example is discussed in Cartwright (1983) – to which the following page numbers refer – pp. 172–4. Cartwright's view is that the collapse of the wave-packet takes place more often than we think. It takes place in particular during the preparation of a system (or ensemble) – see her p. 174. But this makes a nonsense of EPR which mustn't begin with a collapse, if it is to work. But then Cartwright has nothing to say about EPR (see p. 18 of Cartwright 1983).

Chapter 8

An excellent reference on Bell's Theorem is the review article by Clauser and Shimony (1978). The Bell papers (1964) and (1966) are reprinted in Wheeler and Zurek (1983) pp. 397–408.
1 See Margenau (1944).
2 The quotation is from Margenau (1944) p. 188.

3 Eberhard (1977). Eberhard thought his theorem was independent of an assumption of hidden variables, but see Redhead (1983) for a critique of this idea.
4 My treatment follows Redhead (1983).
5 Peierls (1979) p. 28.
6 Bell (1966).
7 Gleason's Theorem is put forward in Gleason (1957).

Chapter 9

1 See Quine (1960) p. 271.
2 Reichenbach (1944). See especially p. 150ff.
3 The classic reference in the literature on quantum logic: Birkhoff and von Neumann 'The logic of quantum mechanics' (1936).
4 See Lemmon (1983).
5 Two such sequent calculi are to be found in Nishimura (1980) which adopts the first interpretation of a sequent, and Cutland and Gibbins (1982) which adopts the second.
6 Lemmon op.cit., pp. 44–9.
7 See Lemmon op.cit., p. 9.
8 For which see Gibbins (1985).
9 Like Ian Hacking in his (1979) paper.

Chapter 10

Hilary Putnam's 1968 paper 'Is logic empirical?' – reprinted as 'The logic of quantum mechanics', in his *Mathematics, Matter and Method: Philosophical Papers Vol. I* (1979) – gets a lot of attention in this chapter. It is a brilliant paper, full of ideas which go off in all directions. We refer to it as 'Putnam (1979)'.

1 Which is Putnam (1979).
2 Putnam (1979) pp. 180–4.
3 Putnam (1979) pp. 180–4.
4 Putnam (1979) p. 180.
5 Putnam (1979) pp. 180–1.
6 Putnam (1979) pp. 181–4.
7 Putnam (op. cit. p. 186): In short, probability (on this view) enters in quantum mechanics just as it entered classical physics, via considering large populations. Whatever problems remain in the analysis of probability, they have nothing special to do with quantum mechanics. But this might not be sufficient to justify the claim that Putnam took quantum probability to be classical. His analysis of the two-slit experiment *does* presuppose classical probability, most obviously in his use of classical conditional probability which Putnam regards as well-defined even for incompatible propositions.
8 Putnam (1979) pp. 184–7
9 Putnam (1979) pp. 183–4 and p. 187.
10 Putnam (1979) p. 185.
11 Putnam (1979) pp. 187–90. The operational meanings of the connectives are considered in pp. 192–7.
12 Putnam (1979) pp. 190–2.
13 The source of this interpretation of Putnam's view on analyticity is his

paper ' "*Two dogmas*" *re-visited*' in his *Realism and Reason: Philosophical Papers Vol III* (1983) pp. 87–97.

14 And it has been so argued. Thus Gardner (1971) pp. 523–4 has two qualitative arguments. First, Gardner argues that the slits prepare 'approximate eigenstates' of position for the incoming particle. These evolve to spread out 'somewhat', but not sufficiently to 'overlap' with the position eigenstate produced by the position measurement at the screen. The trouble with this account is that the evolution in time does more than spread the approximate eigenstates 'somewhat'. In fact, it follows from the total incompatibility of the position operators for different times that they spread out through the whole space instantaneously. Clearly, they spread out sufficiently to 'overlap' with the 'position eigenstate' produced at the screen if a diffraction pattern is to appear at the screen at all. Gardner's conclusion is the correct one. Both sides of the distributive expansion correspond to the zero-vector in Hilbert space and so the distributive expansion is (vacuously) true. But the argument Gardner uses is invalid.

Gardner's second argument is this. Since no source distributes the particles over only a small region of the screen, the set of vectors in Hilbert space which belong to both the subspace spanned by the two eigenstates produced by the slits *and* to any subspace corresponding to a particle's localization to a small region of the screen is empty. This demands a proof. One can be found in Gibbins (1981), and another, more general one in Gibbins and Pearson (1981).

15 The quote is from MacKinnon (1979). Similar remarks are to be found in Levin (1979). I think that there has been a lot of confusion about this. Max Jammer wrote an article claiming that I was wrong in thinking this, but Jammer interestingly misquotes the result, the complementarity theorem when he says that

> no (Lebesque-integrable) function exists which is zero in an interval and whose Fourier coefficients are also zero in an interval. (Jammer 1982, p. 479)

Jammer should have said 'nonzero in a finite interval and whose Fourier coefficients are also nonzero only in a finite interval'. There is an excellent article by P.J. Lahti (1980) which deals with this and other topics, and which includes an illuminating discussion of Bohr's philosophy of quantum mechanics.

16 See Friedman and Glymour (1972).

17 Putnam has explored many quantum logical venues. This one is to be found in Putnam (1974) pp. 60–1.

18 Putnam (1979) p. 189.

19 Putnam (1979) pp. 189–90. Putnam uses '$\wedge$' for our '&'.

20 To be found in Hellman (1980), to my mind the best paper yet written on the meaning of the quantum logical connectives.

21 Hellman (1980) p. 495ff.

22 Bell and Hallett (1982).

23 Bell and Hallett (1982) pp. 363–4.

24 Bell and Hallett (1982) p. 365.

25 Putnam (1979) pp. 192–3. I suspect that few people take the operational definitions of the quantum logical connectives seriously. Of course, begin-

ning with a set of 'tests' is a good way of axiomatizing quantum logic, of trying to recover as much as possible of the mathematical formalism from an austere conception of the theory's empirical basis. It seems that it cannot work entirely. An antioperationist would suspect as much.

26 Putnam (1979) p. 195.

27 Putnam (1979) p. 196.
Later in the chapter I suggest that the meanings of the connectives are parasitic on Hilbert space. Putnam seems to admit as much when he writes (1979) p. 195:

Now *if quantum mechanics is true,* then it turns out that there is an *idealised* test $T_1 \vee T_2$ which is passed by everything that passes T_1 and by everything which passes T_2, and which is such that the things that pass this test pass *every* test such that $T_1 \leq T$ and $T_2 \geq T$.

The first two lots of italics are mine. Assuming 'quantum mechanics is true' subverts the operational definition.

28 This is a stipulation and operationism cannot justify it. Our reason for thinking that there is a test T depends upon our knowledge of the Hilbert space formalism. Dummett has argued that Putnam has no justification for assuming there is such a test in Dummett (1978).

29 On this point see Jauch (1968) pp. 37–9 and Shimony (1971).

30 Dunn (1980).

31 Takeuti (1981).

32 There is an excellent treatment of the Bell inequalities on these lines in d'Espagnat (1979).

33 See Hellman (1981) and Stairs (1982).

34 See Stairs (1982).

References

Ballentine, L. E. (1970) The statistical interpretation of quantum mechanics, *Reviews of Modern Physics,* **42,** 358–81.

Belinfante, F. J. (1975) *Measurement and Time Reversal in Objective Quantum Theory,* Pergamon Press.

Bell, J. S. (1964) On the Einstein–Podolsky–Rosen paradox, *Physics,* **1,** 195–200.

(1966) On the problem of hidden variables in quantum mechanics, *Reviews of Modern Physics,* **38,** 447–52.

Bell, J. and Hallett, M. (1982) Logic, quantum logic and empiricism, *Philosophy of Science,* **49,** 355–79.

Beltrametti, E. G. and van Fraassen, B. C. (eds.) *Current Issues in Quantum Logic,* Plenum Press, New York.

Birkhoff, G. and von Neumann, J. (1936) The logic of quantum mechanics, *Annals of Mathematics,* **37,** 823–43.

Bohm, D. (1951) *Quantum Theory,* Prentice-Hall.

Bub, J. (1979) Conditional probabilities in non-Boolean possibility structures, in *The Logico-Algebraic Approach to Quantum Mechanics,* C. A. Hooker (ed.), Reidel (1979) pp. 209–26.

Cartwright, N. (1983) *How the Laws of Physics Lie,* Oxford University Press.

Clauser, J. F. and Shimony, A. (1978) Bell's Theorem: experimental tests and implications, *Reports on the Progress of Physics,* **41,** 1881–927.

Cutland, N. J. and Gibbins, P. F. (1982) A regular sequent calculus for quantum logic in which $\wedge$ and ν are dual, *Logique et Analyse,* **99,** 221–48.

Daneri, A., Loinger, A. and Prosperi, G. M. (1962) Quantum theory of measurement and ergodicity conditions, *Nuclear Physics,* **33,** 297–312.

DeWitt, B. S. and Graham, N. (1973) *The Many-Worlds Interpretation of Quantum Mechanics,* Princeton University Press.

Dummett, M. A. E. (1978) Is logic empirical? in *Truth and Other Enigmas,* Duckworth, 269–89.

Dunn, J. M. (1980) Quantum mathematics, in *Philosophy of Science Association* (of America) (1980), **2,** 512–31.

Eberhard, P. H. (1977) Bell's theorem without hidden variables, *Il Nuovo Cimento,* **38,** B, 1: 75–9.

Einstein, A., Podolsky, B., Rosen, N. (1935) Can quantum-mechanical description of reality be considered complete? *Physical Review,* Ser. 2, **47,** 777–80.

d'Espagnat, B. (1979) The quantum theory and reality, *Scientific American*, **241**, 128–40.

Feyerabend, P. K. (1968 & 1969) On a recent critique of complementarity, *Philosophy of Science* (I) **35** (1968) 309–31 and (II) **36** (1969) 82–105.

(1975) *Against Method*, New Left Books.

Friedman, M. and Glymour, C. (1972) If quanta had logic, *Journal of Philosophical Logic*, **1**, 16–28.

Gardner, M. (1971) Is quantum logic really logic? *Philosophy of Science*, **38**, 508–29.

Gibbins, P. (1981) A note on Quantum logic and the uncertainty principle, *Philosophy of Science*, **48**, 122–6.

(1985) A user-friendly quantum logic, *Logique et Analyse*, **112**, 353–62.

Gibbins, P. and Pearson, D. (1981) The distributive law in the two-slit experiment, *Foundations of Physics*, **11**, 797–803.

Gleason, A. M. (1957) Measures on the closed subspaces of a Hilbert space, *Journal of Mathematics and Mechanics*, **6**, 885–93.

Hacking, I. (1979) What is logic?, *Journal of Philosophy*, **LXXVI**, 285–319.

(1983) *Representing and Intervening*, Cambridge University Press.

Healey, R. (1985) How many worlds? Noûs, **XVIII**, 591–616.

Heisenberg, W. (1949) *The Physical Principles of Quantum Theory*, Dover Books, New York.

(1971) *Physics and Beyond*, Harper and Row.

Hellman, G. (1980) *Quantum Logic and Meaning*, in Philosophy of Science Association (of America) (1980), **2**, 493–511.

(1981) Quantum logic and the projection postulate, *Philosophy of Science*, **48**, 469–86.

Hooker, C. A. (1972) The nature of quantum mechanical reality: Einstein versus Bohr, in *Paradigms and Paradoxes*, R. G. Colodny (ed.), University of Pittsburgh, 1972, 67–302.

Jammer, M. (1974) *The Philosophy of Quantum Mechanics*, John Wiley.

(1982) A note on Peter Gibbins' "A Note on Quantum logic and the uncertainty principle", *Philosophy of Science*, **49**, 478–9.

Jauch, J. M. (1968) *Foundations of Quantum Mechanics*, Addison-Wesley.

Korner, S. (ed.) (1957) *Observation and Interpretation in the Philosophy of Physics*, Dover Books.

Lahti, P. J. (1980) Uncertainty and complementarity in axiomatic quantum mechanics, *International Journal of Theoretical Physics*, **19**, 789–842.

Laplace (1814) *Essai Philosophique sur les Probabilités*, 2nd edition.

Lemmon, E. J. (1983) *Beginning Logic*, Van Nostrand Reinhold.

Levin, M. E. (1979) Quine view(s) of logical truth, in *Essays on the Philosophy of WV Quine*, R. W. Shahan and C. Swoyer (eds.), University of Oklahoma Press.

Luders, G. (1951) Über die Zustandsanderung durch den Messprozess, *Annalen der Physik*, **8**, 322–8.

Ludwig, G. (1968) *Wave Mechanics*, Pergamon Press.

MacKinnon, E. (1979) Scientific realism: the new debates, *Philosophy of Science*, **46**, 501–32.

Margenau, H. (1944) The exclusion principle and its philosophical importance, *Philosophy of Science*, **11**, 187–208.

McCormmach, R. (1970) The first phase of the Bohr–Einstein dialogue, *Historical Studies in the Physical Sciences,* **2,** 1–39.
(1982) *Night Thoughts of a Classical Physicist,* Harvard University Press.
Mermin, L. D. (1983) The Great Quantum Muddle, *Philosophy of Science,* **50,** 651–6.
Newton, I. (1952) *The Opticks,* Dover Publications.
Nishimura, H. (1980) Sequential method in quantum logic, *Journal of Symbolic Logic,* **45,** 339–52.
Peacock, G. (1855) *Miscellaneous Works of the Late Thomas Young MD FRS &c,* John Murray, London.
Peierls, R. (1979) *Surprises in Theoretical Physics,* Princeton University Press.
Popper, K. R. (1967) Quantum Mechanics without 'The Observer', in *Quantum Theory and Reality,* M. Bunge (ed.), Springer-Verlag, 1967, pp. 7–44.
Putnam, H. (1974) How to think quantum-logically, *Synthese,* **29,** 55–61.
(1979) The logic of quantum mechanics, in his *Mathematics, Matter and Method: Philosophical Papers Vol I,* Cambridge University Press, 174–97.
(1983) 'Two dogmas' re-visited, in his *Realism and Reason: Philosophical Papers Vol III,* Cambridge University Press, 87–97.
Quine, W. V. O. (1960) *Word and Object,* MIT Press.
Raman, V. V. and Forman, P. (1969) Why was it Schrödinger who developed de Broglie's ideas? in *Historical Studies in the Physical Sciences 1,* University of Pennsylvania Press, 1969, 291–314.
Redhead, M. L. G. (1982) Quantum field theory for philosophers, *Philosophy of Science Association* (of America) (1982) **2,** 57–99.
(1983) Nonlocality and peaceful coexistence, in *Space, Time and Causality,* R. Swinburne (ed.), Reidel.
Reichenbach, H. (1944) *Philosophic Foundations of Quantum Mechanics,* University of California Press.
Rosa, R. (1979) Electron interference: Lande's approach upset by a recent elegant experiment, *Nuovo Cimento Letters,* **24,** 549–50.
Rozenthal, S. (1967) *Niels Bohr,* North-Holland.
Scheibe, E. (1973) *The Logical Analysis of Quantum Mechanics,* Pergamon.
Shimony, A. (1971) Filters with infinitely many components, *Foundations of Physics,* **1,** 325–8.
Stairs, A. (1982) Quantum logic and the Luders rule, *Philosophy of Science,* **49,** 422–36.
Stuewer, R. H. (1975) *The Compton Effect: Turning Point in Physics,* Science History Publications, New York.
Takeuti, G. (1981) Quantum set theory, in *Current Issues in Quantum Logic,* Beltrametti, E.G. and van Fraassen, B.C. (eds.) Plenum Press, New York, 1972, 302–22.
Taylor, G. I. (1971) *The Scientific Papers of Sir Geoffrey Ingram Taylor Vol. IV,* G. K. Batchelor (ed.), Cambridge University Press.
van der Waerden, B. L. (1967) *Sources of Quantum Mechanics,* Dover Publications.
van Fraassen, B. C. (1970) On the extension of Beth's semantics of physical theories, *Philosophy of Science,* **37,** 325–39.

von Neumann, J. (1955) *Mathematical Foundations of Quantum Mechanics,* Princeton University Press.

Wheaton, B. R. (1983) *The Tiger and the Shark: Empirical Roots of Wave–Particle Duality,* Cambridge University Press.

Wheeler, J. A. and Zurek, W. H. (1983) *Quantum Theory and Measurement,* Princeton University Press.

Wigner, E. (1961) Remarks on the mind–body question, in *The Scientist Speculates,* I. J. Good (ed.), Basic Books, 1961, 284–302.

Wittgenstein, L. (1968) *Philosophical Investigations,* Basil Blackwell.

Wu, C. S. and Shaknov, I. (1950) The angular correlation of scattered radiation, *Physical Review,* **77,** 136.

Index

a prioricity, 146
Absolute Space, 4
analyticity, 146
Arago D, 42

Ballentine LE, 77, 78, 80, 114
Belinfante FJ, 76
Bell JL, 156–7
Bell JS, 11, 118
Bell inequalities, 118–22, 151, 161–2
Birkhoff G, 127
black-body radiation, 19, 22
Bohm D, 73
Bohr N, 6, 9, 15, 22, 24, 42, 44, 47, 49, 52, 60, 62, 80, 88, 109, 165
  Bohr's Como lecture, 56, 65
Bohr–Einstein dialogue, 11, 62–75
Bohr–Sommerfeld theory of the atom, 22, 50
Bohr–Kramers–Slater theory, 63–4
Born M, 28
Born's particle ontology, 45–6
Bothe–Geiger experiment, 64, 65
Boyle R, 3
Bruno G, 5

Carnap R, 127
Cartwright N, 15, 113–114
classical concepts, priority of, 49, 53
collapse of the wave-packet, 11, 39
commutator, 26
complementarity, 50, 53ff
complementarity theorem, 27, 45
Compton AH, 23
  Compton effect, 23, 43, 51, 63
consciousness, 12
Copenhagen interpretation, 9, 11, 41, 47–61, 62–83, 165
correspondence principle, 23
  generalised, 29

Daneri–Loinger–Prosperi theory of measurement, 113
Davisson CJ, 39
de Broglie, 23, 24, 43
delayed-choice experiment, 40–1
Democritean atomism, 3
  ontology, 15
density operator, 98–100
Descartes R, 5
Dirac notation, 32
Duane W, 78, 82
Dunn JM, 160

Eberhard PH, 118
Ehrenfest P, 20
eigenvalue–eigenstate link, 30
Einstein A, 6, 9, 21, 22, 43, 46, 54, 62, 66
  photon-in-a-box thought-experiment, 68–70
Einstein–Podolsky–Rosen thought-experiment, 11, 13, 47, 71–5, 114
  necessary condition for completeness, 72
  sufficient condition for 'element of reality', 72
  version due to Bohm, 73, 119
ELQM, the elementary language of quantum mechanics, 93ff
ensembles, 9
expectation-value, 31

Feyerabend PK, 81, 88
Fresnel A, 42
Friedman M, 152

Germer LH, 39
Gleason AM, 124
Glymour C, 152

Hacking I, 15
Hallett M, 156–7
Hardegree H, 138
Heisenberg W, 7, 9, 24, 44, 47, 49, 50, 52, 80
  Chicago lectures, 59
  gamma-ray microscope, 50–3, 103
  matrix mechanics, 23
Hellman G, 155, 163
Hertz H, 2
hidden variables, 9, 116–25
Hilbert space, 87
  subspaces, 92
  tensor product, 101, 110
Huyghens C, 42

ideal measurements, *see* measurements of the first kind
indeterminacy relations, 10, 27, 50, 54, 59–61, 66
interference, 46
  constructive, 38
  destructive, 38, 149
  self, 38–40

kinetic theory of gases, 8
Kochen–Specker theorem, 124

Laplace P de, 4
lattice, 92, 128ff
  modular, 96
  orthomodular, 97
Leibniz–Clarke correspondence, 62
Lemmon EJ, 131, 137, 140
Lewis GN, 43
locality, 71, 116
Locke, 3
Lummer and Pringsheim measurements, 20, 21

Mach E, 2, 7
MacKinnon E, 149–50
many-worlds interpretation of quantum mechanics, 111–12
Margenau H, 117
Maxwell JC, 2, 3, 5
McCormmach R, 23
measurements
  of the first kind, 105
  maximal and nonmaximal, 106
Mermin LD, 82
meta-physics, 1
Millikan RA, 42
minimal ensemble interpretation, 76, 77
mixtures, 97–8
  ignorance interpretation of, 114–15
  proper and improper, 99
Moller PM, 55

Newton I, 3, 4, 5, 40, 41, 62
newtonian world-picture, 3
no hidden-variables theorems, 122–5
nonlocality, 12, 116–25

old quantum theory, the, 7
operators, 25
  Hermitian, 91
  projection, 92
  trace of, 99–100

Pauli, 33
  exclusion principle, 117–18
Phenomenon, the, 56, 58–9, 75
photo-electric effect, 21, 22, 36
picturing, 58
Planck M, 6, 20, 21, 24
  Planck's law, 21
polarization of light, 35
Popper KR, 45, 77, 78, 79–82, 115
probability, 8, 75
projection postulates, 102–15
  the clumsy experimenter's rule, 107, 163
  Luders' rule, 108, 163, 165
Putnam H, 141, 143ff

quantum logic, 91
  conditional in, 138
  conditional probability on, 147, 150
  as a lattice, 95–7
  meanings of the connectives, 153–9
  natural deduction rules of inference, 132, 134–5
  nonmodularity, 91
quantum logical interpretations
  activist, 142
  passivist, 142
  preservationist, 143
  quietist, 142, 160, 167
  realist, 151–3
  revisionist, 143
quantum postulate, 50, 53, 54, 55
Quine WVO, 127, 146

relativity, 6
Robertson HP, 59
Rontgen W, 42
Rosenfeld L, 48

Rutherford W, 22
Ryle G, 154

Schrodinger E, 7, 24, 28, 44, 54
Schrodinger's wave mechanics, 23, 43
wave ontology, 43
Schrodinger's Cat, 76, 111, 113
semantic ascent, 127
Solvay Congress
1921, 63, 66
1927, 64–5
1930, 65, 68, 70
space-quantisation, 33
spin matrices, 33
Stairs A, 163
state-vector, 25
Stern-Gerlach effect and experiment, 32, 103, 104
statistical mechanics, 8, 20
suprerposition principle, 27, 28

Takeuti G, 160
Taylor Sir GI, 39, 42
Thomson JJ, 42
two-slit experiment, 13, 36, 41, 117, 147–51

ultraviolet catastrophe, 20
uncertainty principle, *see* indeterminacy relations

von Neumann J, 88, 109, 122, 127, 163, 165

wave–particle duality, 27, 36–46
well-formed-formulae, 131–2
Wheeler JA, 40
Wittgenstein L, 48, 58

Young T, 37